Introductory Quantitative Analysis

Larry E. Wilson
Ohio University–Lancaster

Charles E. Merrill Publishing Company
A Bell & Howell Company
Columbus, Ohio

To the enlightenment of
my students—past, present,
and future

Published by
Charles E. Merrill Publishing Co.
A Bell & Howell Company
Columbus, Ohio 43216

International Standard Book Number: 0-675-08869-0

Library of Congress Catalog Number: 73-87528

2 3 4 5 6—79 78 77 76 75

Printed in the United States of America

Preface

This text is designed as the basis for a beginning course in quantitative analysis—either near the end of the freshman year or during the sophomore year in college. The scope and depth of the material should fit nicely into a one-semester course. By abbreviating the material somewhat with selective omissions, the text could be used in a one-quarter course taught during the freshman year. It is assumed that the course using this text will serve only as an introduction to analytical chemistry and that nearly all the students will take additional analytical courses which cover instrumental methods of analysis in much more detail.

The major purpose of this text is to provide the student with an adequate background of the basic fundamentals and techniques of classical quantitative analysis and to present this material in a manner which can be readily comprehended. Since most beginning quantitative analysis courses utilize a pH meter and a simple spectrophotometer, the theory of these two instruments along with some experimental procedures is included.

It is felt that an early introduction into the theory and especially the laboratory techniques of quantitative chemical analysis is important in

iii

the training of scientists and health-related professional people. There is nothing that is as relevant in the training of a scientist or a health-related professional as a systematic approach to the problem at hand and an application of extreme care while at work. Both of these attributes are stressed in classical wet-chemical methods.

Problems are listed after the chapters to enable the student to evaluate his comprehension of the material. Answers to all numerical problems are given in Appendix C. The student should attempt to fully understand each problem and not just work towards the answer. The formula weights of all compounds involved in the numerical problems and all reagents used in the laboratory procedures are listed in Appendix B. A two-place logarithm table is provided in Appendix B to aid the student in making pH calculations. A table listing the properties of concentrated acids and bases is also located in Appendix B for reference in preparing standard solutions.

Laboratory procedures are given in each area covered in the text: gravimetric analysis; volumetric analysis, including acid-base, redox, precipitation, and complexometric titrations; and spectrophotometric methods of analysis. A sufficient number of experiments are presented to allow the instructor to use those which he finds most helpful and necessary to each individual class or student.

I wish to express my appreciation to my students who have used this textbook in its incomplete developmental stages. Special thanks go to Danny Jenkins, Jane Good, and Beverly Groves, undergraduate laboratory assistants, who have helped proofread the manuscript.

Contents

An Introduction to Analytical Chemistry

Analytical chemistry consists of two major divisions, qualitative analysis and quantitative analysis. Qualitative procedures are designed to identify *what* is present in a sample, whereas the quantitative procedures are designed to determine *how much* of a substance is present in a sample. A complete analysis of a sample includes both qualitative and quantitative determinations. This book deals only with quantitative analysis. It is presumed that previous qualitative determinations have been made as they indeed form the basis for selecting quantitative procedures.

Divisions of Quantitative Analysis

Quantitative analysis procedures may be subdivided into three principle divisions: gravimetric analysis, volumetric (or titrimetric) analysis, and instrumental analysis. Instrumental analysis may be further subdivided into those procedures involving optical methods and those involving electrical methods. As in any area of science, absolute delineation of these divisions is impossible. Many methods involve two or more of these areas. The above division of methods is based upon the principle type of measurement made in the analysis.

1. Gravimetric analysis

A gravimetric analysis is based entirely on weight. The original sample is weighed; and the substance to be determined or a known derivative of the substance is isolated and also weighed. From the two weights, the percent of the substance is calculated.

2. Volumetric analysis

The volume of a solution with a known concentration, which reacts in a definite manner with the substance being determined, is measured. From the original sample weight, and the reacting strength of the solution, the percent of the substance is calculated.

3. Instrumental analysis

Most instrumental methods of analysis can be classified in one of the following categories:

    A.   Optical methods: Optical methods of analysis involve the measurements of the wavelength or intensity of radiant energy.

    B.   Electrical methods: Electrical methods of analysis involve the measurements of one or more fundamental electrical quantity such as voltage, resistance, or current.

This text is primarily concerned with gravimetric analysis and volumetric analysis, which are sometimes referred to as classical or wet-chemical analytical methods.

What Is Required in an Analysis?

Before undertaking a quantitative analysis of a sample, certain questions must be answered.

What is to be determined? Rarely is a complete analysis required. Usually only one or two constituents will be determined. Sometimes, e.g., in assays, only the major constituent is determined. In other cases, only the determination of trace impurities is required.

How much accuracy is required? Depending on the end use of the sample, different degrees of accuracy may be required. Time, effort, and money are wasted doing a highly accurate but tedious determination if a less accurate answer is perfectly satisfactory and can be obtained more easily.

What possible interferences are there? To answer this question some history of the sample must be known. Perhaps preliminary separations or the addition of a reagent which prevents a species from interfering will be required.

Only after the above questions have been answered can the analyst decide which method to employ in the analysis. Fortunately for the student in a beginning course in quantitative analysis, this decision has usually been made. Often no interferences are present and the percent of a major constituent is determined by a standard procedure.

Methods of Expressing Concentrations of Solutions

The fundamental unit which measures the quantity of substance is the mole. A mole is defined as Avogadro's number (6.023×10^{23}) of molecules of that substance. Chemists use the term mole in its broad sense to describe the amounts of molecular compounds, free elements, ions, and chemical changes. A mole expresses in grams the sum of the atomic weights of a molecular compound, ionic compound, element, or ion.

$$1 \text{ mole of } H_2O = 18.01 \text{ grams}$$

$$1 \text{ mole of } NaCl = 58.44 \text{ grams}$$

$$1 \text{ mole of } Fe = 55.85 \text{ grams}$$

$$1 \text{ mole of } Cl^- = 35.45 \text{ grams}$$

Many different methods are used to express the concentration of a solute in a solution. These methods include the following:

1. Molarity. The molarity of a solution expresses the number of moles of solute in one liter of solution.

2. Molality. The molality of a solution expresses the number of moles of solute in one kilogram of solvent.
3. Normality. The normality of a solution expresses the number of equivalents of solute in one liter of solution.
4. Formality. The formality of a solution expresses the number of formula weights in one liter of solution.
5. Mole fraction. The mole fraction of a solution is the ratio of the number of moles of solute to the total moles of solute plus solvent. This method is not often used in analytical chemistry.
6. Weight percent. The weight percent of a solution is the number of grams of solute in 100 grams of solution.
7. Volume percent. The volume percent of a solution is the number of milliliters of solute in 100 milliliters of solution.
8. Parts per million. The concentration of trace constituents is often expressed in parts per million. Strictly speaking, parts per million expresses the number of milligrams of solute in a kilogram of solution. This expression is often applied to dilute aqueous solutions where one liter of solution weighs approximately one kilogram. Parts per million, therefore, is generally used to express the number of milligrams of solute in one liter of solution.

Methods 1–5 are used to express exact analytical concentrations.

Analytical concentrations are not to be confused with equilibrium concentrations which are required for equilibrium considerations. For example a $1.00M$ NaCl solution is one which is prepared by dissolving 58.44 g of sodium chloride in water and diluting the solution to one liter. Sodium chloride is 100 percent ionized in solution, so the equilibrium concentration of NaCl is essentially nil. The equilibrium concentration of Na^+ and Cl^- is $1.0M$. A $0.1000M$ acetic acid is a solution containing 6.005 g of CH_3COOH in a liter of solution. As will be seen in Chapter 7, acetic acid is a weak acid and is slightly ionized in an aqueous solution. The actual equilibrium concentration of molecular acetic acid is $0.0988M$. For exact calculations involving equilibrium, equilibrium concentrations, not analytical concentrations, are required. Because of the possible implications involved by expressing $1.00M$ NaCl and $0.1000M$ CH_3COOH, many chemists have resorted almost exclusively to formality to express concentrations. Formality expresses the number of formula weights per liter disregarding any ionizations, dissociations, or associations involved.

Brackets are used to signify equilibrium molar concentrations. For the $0.1000M$ CH_3COOH solution described above, the following concentrations exist: $[CH_3COOH] = 0.0988$, $[H^+] = 1.2 \times 10^{-3}$, $[CH_3COO^-] = 1.2 \times 10^{-3}$. These mean that there are 0.0988 moles of nonionized acetic acid and 1.2×10^{-3} moles of hydrogen ions and of acetate ions per liter.

Methods 6–8 may be used to express exact concentrations but are commonly used to express approximate concentrations. A weight percent solution is generally prepared by weighing the solute approximately on a rough balance and adding it to water measured with a graduated cylinder. For example, a 10% barium chloride solution being 10 g $BaCl_2$ in 90 ml water (weighing approximately 90 g). A volume percent solution is always prepared with the solute and solvent measured with a graduated cylinder. Often a cross between weight percent and volume percent is used. A 5% glucose solution may be 5 g of glucose and 100 ml (≈ 100 g) of water (or solution). Since these concentrations are only approximate, directions for preparation of these types of solutions are often lacking. If the method calls for a 10% $BaCl_2$ solution, little matters if the solution is prepared by adding 10 grams of $BaCl_2$ to 90 g, 100 g, or 100 ml of water.

Table 1-1 lists the different methods for expressing concentration along with standard notations and units. A basic knowledge of units is necessary in analytical chemistry.

Table 1-1

Methods of Expressing Concentrations of Solutions

Method	Notation	Units
Molarity	M	moles/liter
Molality	m	moles/kilogram
Normality	N	equivalents/liter
Formality	F	formula weights/liter
Mole fraction	χ	unitless
Weight percent	wt %	unitless
Volume percent	vol %	unitless
Parts per million	ppm	milligrams/kilogram or milligrams/liter

Activity and Ionic Strength

The molar concentration of a substance in solution can be accurately known by the method of preparation or it can be determined by a gravimetric or volumetric determination. This concentration is also known as the analytical concentration. In a very dilute solution of a strong electrolyte, relatively few positive and negative ions are available, and attractive and repulsive interactions are at a minimum. As the concentration of ions in a solution is increased, the extent of attraction and repulsion becomes more significant. The free motion of the ions is impeded and the ions are not as active as free ions. The net effect is that the *effective* concentration of ions in the solution is not as great as the actual analytical concentration of ions.

This difference between the effective concentration and the actual analytical concentration becomes greater as the concentration increases. The effective ion concentration can be determined by conductivity measurements, specific ion electrode measurements, e.g., pH meter, and other methods. The term *activity* is used to denote the effective concentration of the substance in solution. The relation between concentration and activity is stated mathematically as follows:

$$a = fc$$

or

$$a = f[\;\;]$$

Where a is the activity, c or [ ] is the molar concentration, and f is the *activity coefficient*.

The activity coefficient depends on the ionic strength of the solution, which is defined mathematically as follows:

$$\mu = \frac{1}{2}\sum [i]Z_i^2$$

Where μ is the ionic strength, $[i]$ is the molar concentration of an ion, and Z_i is the charge on the ion.

The Debye-Hückel limiting law equation relates the activity coefficient of an ion to the ionic strength of the solution. The simplified Debye-Hückel equation is

$$-\log f = AZ_i^2 \sqrt{\mu}$$

Where A is a constant relating the dielectric constant of the solution with the closeness of approach of the ions. The value of A for dilute aqueous solutions is equal to 0.505. The Debye-Hückel equation should not be applied to solutions which are much more concentrated than $0.01\,M$. An extended equation is applicable up to $0.5\,M$. For more concentrated solutions, electrode potentials, freezing-point depressions, and solubility measurements are usually used.

In dilute solutions activity coefficients approach unity. However, they decrease in value as the ionic strength of the solution increases. Equilibrium constants are strictly constant only at infinite dilutions. Table 1-2 shows the molar solubility of AgCl in KNO_3 solutions.

Table 1-2

Solubility of Silver Chloride in Dilute Potassium
Nitrate Solutions at 25°C
(S_0 = molar solubility in pure water; S = molar
solubility in electrolyte solution)

Concentration of KNO_3, M	Solubility AgCl	S/S_0
0	1.278×10^{-5}	1.00
0.001	1.325×10^{-5}	1.04
0.005	1.385×10^{-5}	1.08
0.01	1.427×10^{-5}	1.12

Source: S. Popoff and E. W. Neuman, *J Phys. Chem.* **34**, 1853 (1930).

Table 1-2 clearly shows that activities should be considered in rigorous calculations involving equilibrium constants. In this textbook activity coefficients are ignored. The preceding information has been presented to make the student aware of the theoretical considerations involved in exact equilibria calculations.

Questions

1. Define the following terms: gravimetric analysis, volumetric analysis, instrumental analysis, mole, molarity, normality, mole fraction, activity, ionic strength, activity coefficient, Debye-Hückel equation.

2. Discuss the difference between the analytical concentration of a species and the equilibrium concentration of a species.

3. Why would the directions, "Prepare a 5% aqueous solution of sodium hydroxide," be vague?

4. What physical property of a solution must be known in order to convert molarity to molality?

5. Explain why the number of milligrams of solute in a dilute aqueous solution is said to equal the concentration in parts per million.

Problems

1. Calculate the molecular weight (to the nearest 0.01 g) of each of the following compounds.
   (a) Ag_2S (b) $LiCl$
   (c) $CHCl_3$ (d) UF_6
   (e) $Pb(C_2H_3O_2)_2$ (f) $NaMg[UO_2(C_2H_3O_2)_3]_3 \cdot 5H_2O$

2. Calculate the weight in grams of each of the following:
   (a) 2.5 moles of $AgNO_3$ (b) 1.2 moles of ethyl alcohol (CH_3CH_2OH)
   (c) 3.2 moles of Cl^- (d) 2.0 moles of Cu

3. Calculate the molarity of a solution prepared by dissolving 150 g of $NaCl$ in water and diluting it to 1.000 liter.

4. Calculate the weight of Na_2SO_4 which must be dissolved in 250 ml of water (a) to give a solution which is $0.20M$ and (b) to give a solution which is $0.20M$ in sodium ion.

5. Calculate the number of moles of barium chloride, $BaCl_2$, in 150 ml of $0.22M$ solution. Assuming the salt to be completely ionized in solution calculate the molar concentration of barium ion and chloride ion in the solution.

6. Calculate the number of moles present in the following:
   (a) 454 g $AgNO_3$ (b) 25 g $KMnO_4$
   (c) 12.5 g $NaCl$ (d) 50 ml of $0.10M$ $AgNO_3$
   (e) 250 ml of $2.0M$ H_2SO_4 (f) 1.00 ml of $0.01M$ $KSCN$

7. Calculate the approximate molarity of a 5.5% (weight-volume) solution of $BaCl_2 \cdot 2H_2O$ in water.

8. Calculate the approximate number of moles of potassium hydroxide in 25 ml of a 10% (weight-volume) solution of KOH in ethanol.

9. Calculate the molarity of a solution of concentrated hydrochloric acid having a density of 1.18 g/ml containing 36% HCl by weight.

10. The equivalent weight of $KMnO_4$ in an acid solution is one-fifth its molecular weight. Calculate the normality of a solution prepared by dissolving 1.62 g $KMnO_4$ in water and diluting to 500 ml.

11. Calculate (a) the molarity of sodium chloride in a solution prepared by the mixing of 50.0 ml of $0.100M$ $NaCl$, 25.0 ml of $0.100M$ HCl, and 75 ml of water and (b) the molarity of the chloride ion in the solution.

12. Calculate the normality of a solution prepared by diluting 5.00 ml of $2.000N$ $NaOH$ to exactly 100.0 ml.

13. Calculate the volume that 50 ml of $16M$ HNO_3 should be diluted to give a $0.5M$ solution.

14. A $0.20M$ solution of formic acid is 3.2% ionized. Calculate the equilibrium concentration of formic acid and the formate ion in a $0.20M$ solution.

15. Calculate the concentration in parts per million of dissolved oxygen in a stream if a 500 ml sample contains 25 ml of O_2. A mole of oxygen gas occupies 22,400 ml.

16. Calculate the weight of dissolved calcium carbonate in a 250 ml sample of hard water containing 250 parts per million calcium carbonate.

17. Calculate the ionic strength of the following solutions:
    (a) $0.1M$ NaCl   (b) $0.05M$ $BaCl_2$
    (c) $0.02M$ $Al_2(SO_4)_3$ (d) $0.005M$ Th $(NO_3)_4$

18. Calculate the ionic strength of a solution containing $0.05M$ KNO_3 and $0.01M$ K_2SO_4.

19. Using the simplified Debye-Hückel equation calculate the activity co-efficient of the following ions:
    (a) Na^+ at $\mu = 0.01$ (b) Ba^{++} at $\mu = 0.05$
    (c) Al^{+++} at $\mu = 0.0003$ (d) Pb^{++} at $\mu = 0.10$

20. Calculate the activity of the calcium ion in a $0.02M$ solution of $CaCl_2$.

2

Operations of Quantitative Analysis

It is the purpose of this chapter to provide a general introduction to the laboratory operations which are common to most quantitative analysis procedures.

Laboratory Notebook

A permanently bound notebook is used for the recording of all data. The data is recorded *in ink* at the time it is observed. Under no conditions is the use of pencil or loose scratch paper tolerated in a quantitative analysis laboratory.

Orderliness in a notebook is advantageous. A little time spent before the laboratory period in planning and organization is time well spent. Data should never be recorded in one portion of the notebook and later transferred to the proper data page. This would introduce the chance of transposition of numbers. In recording data, there is always the possibility of writing down the data incorrectly. Before readjusting the balance, buret, or whatever the source of data, you should check to make sure that the data are correct. If an error has been made, the incorrect data are crossed out with a *single line* and the correct data are written above. Never erase data nor cross them out so as to obliterate them. It is to be expected in the course of a term's work that some data will be incorrectly recorded. It does not detract from the neatness of the notebook if the mistaken data are corrected as described above. Never for any reason is a page removed from the laboratory notebook.

The pages of the bound notebook should be numbered. The first few pages should be reserved for a table of contents and this should be kept up to date as work progresses. Each page should be dated. The name of the determination should appear at the top of the first page on which the experimental data is recorded. Figure 2-1 shows a form for a typical gravimetric determination. Figure 2-2 shows a form for a typical

Chloride Determination

Date   4-8-71 and 4-15-71   Unknown no.   495

Beaker number	1	2	3
Wt of W.B. + sample	30.5452	30.0377	29.5674
Wt of W.B.	30.0377	29.5674	29.0985
Wt of sample	0.5075	0.4703	0.4689
Crucible number	I	II	III
Wt of crucible plus ppt	14.0972	14.4120	15.0900
Wt of crucible	13.0108	13.4050	14.0858
Wt of ppt	1.0864	1.0070	1.0042
% of Cl$^-$	52.96	52.97	52.98
Average %		52.97	
Range		0.02%	
Standard deviation		0.01%	
Chemical equation			

$$Cl^- + Ag^+ \rightarrow AgCl$$

Math equation

$$\%Cl^- = \frac{\text{wt of ppt} \times \dfrac{Cl}{AgCl} \times 100}{\text{sample wt}}$$

Figure 2-1

Form for gravimetric determination.

Standardization of NaOH

Date 4-15-71

Run	1	2	3
Wt of W.B. + KHP	39.3584	38.5463	37.7305
Wt of W.B.	38.5463	37.7305	36.9167
Wt of KHP	0.8121	0.8158	0.8138
Volume of NaOH	32.62	32.78	32.68
Normality of NaOH	0.1219_0	0.1218_6	0.1219_3
Average normality		0.1219	

Determination of KHP

Unknown no. 160

Run	1	2	3
Wt of W.B. + sample	23.8161	21.8039	19.7687
Wt of W.B.	21.8039	19.7687	17.7218
Wt of sample	2.0122	2.0352	2.0469
Volume of NaOH	28.72	29.08	29.25
% of KHP	35.53	35.57	35.58
Average %		35.56	
Range		0.05%	
Standard deviation		0.03%	

Chemical equation

$$HC_8H_4O_4^- + OH^- \rightarrow C_8H_4O_4^{-2} + H_2O$$

Math equations

$$\text{normality NaOH} = \frac{\text{sample wt}}{\dfrac{\text{KHP}}{1000} \times V_{ml}}$$

$$\%\text{KHP} = \frac{V_{NaOH} \times N_{NaOH} \times \dfrac{\text{KHP}}{1000} \times 100}{\text{sample wt}}$$

Figure 2-2

Form for volumetric determination.

volumetric determination. Note that it is often necessary to use standard abbreviations in the recording of data. Wt refers to weight, W.B. to weighing bottle, ppt to precipitate, and KHP to potassium acid phthalate. The subscript notation on the normality will be covered in the discussion of significant figures in Chapter 3.

In addition to the information recorded in Figures 2-1 and 2-2, complete calculations including statistical treatment should be entered into the notebook. Calculations should be done with a calculator or a logarithm table. Slide rule accuracy is not adequate. Longhand calculations to four significant places often lead to errors. If a printing calculator is available, the tape of the calculations with appropriate identification can be fastened to the notebook with tape. It is convenient to use the right-hand pages of the notebook for the recording of data and the left-hand pages for calculations.

The basic form for recording data should be placed in the notebook prior to the beginning of the laboratory period. Without this form, careless mistakes can be made resulting in time-consuming and otherwise needless repetition of work. With the form before him, an analyst is not likely to use a crucible without obtaining and recording its weight or to neglect to measure and record some other equally important measurement.

Occasionally it may occur that a complete page of data becomes useless. An example would be in a volumetric determination where it was discovered after the titration that some contamination of the titrant had occurred which could not be corrected by restandardization. In this case a single x drawn through the page with an accompanying explanation is all that is necessary to void the data. It is not necessary to cross out each data entry, and never should the page be removed nor the data obliterated.

Laboratory Report

Usually in a course in quantitative analysis, a laboratory report is given to the instructor after the work is finished, and the instructor grades the report according to the amount of accuracy of the determination. Normally, grading is based on the average of a series of determinations, with doubtful values not being included in the average. Statistical consideration (see Chapter 3) can be applied to eliminate doubtful values from the average. The instructor may require that at least two determinations fall within a certain range in order for a report to be acceptable. Additional determinations may be necessary to obtain the necessary range and average.

The laboratory report should be written in ink. It should include all data and results as entered in the notebook. A Xerox copy of the notebook page is acceptable. The laboratory report should include all pertinent chemical equations and a word expression of any mathematical equations involved. The data for one determination should be substituted into the mathematical equation as an example. The analyst's name should be placed at the top of the report and his signature at the end.

Desiccator

A desiccator is an airtight container which maintains an atmosphere of low humidity when the bottom portion is filled with a suitable drying agent. Figure 2-3 shows a typical desiccator used in a beginning course in quantitative analysis. A desiccator is used both for the cooling of heated objects and for the storage of dry objects in an atmosphere of controlled humidity. Some common desiccants and their comparative drying powers are given in Table 2-1.

Figure 2-3

Desiccator for cooling samples.

Table 2-1

Drying Powers of Some Desiccants

Desiccant	Residual water in mg per liter of air at 30.5°C
BaO	0.0007
$Mg(ClO_4)_2$, anhydrous	0.002
H_2SO_4, concentrated	0.003
$CaSO_4$, anhydrous	0.005
Al_2O_3	0.005
Silica gel	0.03
$CaCl_2$, anhydrous	0.36

Anhydrous calcium sulfate sold by W. A. Hammond Drierite Co. under the trade name of Drierite is a powerful drying agent. An indicating Drierite is also available which is anhydrous calcium sulfate coated with a layer of cobalt sulfate. When dry, the indicating Drierite is blue. When the Drierite turns pink the desiccant should be replaced. Only a small percentage of the indicating Drierite needs to be mixed with regular Drierite.

An airtight seal is maintained by application of stopcock grease to the ground glass rim on the top of the desiccator. Once the desiccant has been added and the stopcock grease applied, the top of the desiccator should not be removed any more than necessary because this will allow moist air to cause an unnecessary demand on the desiccant. An ignited crucible or other very hot object should be cooled for about 30 seconds before being placed in the desiccator. Objects dried at 100°–275°C can be placed directly into the desiccator. Under either condition, the lid of the desiccator should be slightly ajar for 30 seconds prior to complete closing. This is to prevent a partial vacuum from forming as the heated air cools. If a partial vacuum would form, two problems might arise. First, the lid may be difficult to remove, which could result in the samples becoming upset when the lid is forced open. Second, the inrush of air into the evacuated desiccator can sweep some of the sample from its container. Consequently, when removing the lid, proper care must be exercised. A dried sample should remain in the desiccator for at least 30 minutes before being weighed. If the sample is to remain in the desiccator for an extended period of time before weighing, the top of the weighing bottle should be put in place *after* 30 minutes cooling.

Weighing bottles and crucibles should not be handled directly with the fingers after drying. A 1.5–2 inch piece of 3/16 in I.D. rubber tubing slit lengthwise provides an effective but inexpensive finger cot for handling dried glassware and crucibles for weighing. Figure 2-4 shows the appropriate technique. Thin cotton gloves can also be used for handling glassware. Crucible tongs should be used if the equipment is too hot to handle.

Figure 2-4

Rubber finger cots for handling object being weighed.

Drying a Sample

If a sample is to be dried before weighing, transfer it into a clean weighing bottle. Place the weighing bottle and lid (removed) in a clean beaker. Cover the beaker with a ribbed watchglass or a regular watchglass supported on glass hooks (Figure 2-5). Place the beaker and contents in the oven for the specified length of time. After drying, transfer the bottle and lid to the desiccator and cool as described in the previous section. Do not put the lid of the weighing bottle in place until the 30 minute cooling period has elapsed.

Weighing with an Analytical Balance

There is a wide variation of analytical balances in use in laboratories of quantitative analysis, but all fall into one of two categories. The

Figure 2-5

Technique for drying in oven.

conventional double-pan balance has been widely replaced with the more rapid single-pan substitution type balance with readout device. The single-pan balances are much easier to use, with a typical weighing being completed in about a minute. Single-pan balances, however, are not necessarily more accurate or precise than double-pan balances.

Both single-pan and double-pan balances are based upon the same principles. The analytical balance is a first class lever that compares the mass of an unknown object with the mass of standard objects known as weights. The terms mass and weight are used interchangeably in quantitative analysis.

A double-pan balance (Figure 2-6) consists of a strong but light weight beam containing three prism-shaped agate knife edges located in a plane, with the two terminal ones equidistant from the center knife edge. The terminal knife edges are inverted from the center knife edge and support the stirrups, which in turn support the pans on which the object and weights are placed. The center knife edge lies slightly above the center of gravity of the beam and pan assembly and rests on a smooth agate plate located on a center post. The pointer extends from below the center knife edge to a pointer scale mounted at the bottom of the post. The object is placed on the left pan and the weights are placed on the right pan. The point of equilibrium attained with the empty balance is restored when the weight of the weights equals the

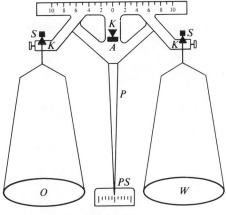

Figure 2-6

Double-pan analytical balance showing
the three knife edges (*K*), agate plate (*A*),
stirrups (*S*), pointer (*P*), pointer scale (*PS*),
object pan (*O*), and weights pan (*W*).

weight of the object. Standard methods are used for determining the
point of equilibrium of a pointer which is swinging back and forth.
Consult with your instructor for the method desired if you use a
double-pan balance.

A single-pan balance (Figures 2-7 and 2-8) operates with a beam
under a constant load. These balances contain a beam with one end
that holds a pan and a series of weights and the other end is counter-
poised by a single weight. When an object is placed on the pan,
weights are removed from the beam to restore it to equilibrium.
Actually, the beam is not brought completely into balance. Weights
are removed only to the nearest 0.1 g or 1 g. The imbalance of the
beam is registered optically and automatically on an illuminated vernier
scale (Figure 2-9). The weights of a single-pan balance are removed
by control knobs on the front of the balance and are registered on a
counter also located on the front.

Weighing the Sample

Usually three or more samples are weighed for a quantitative determination. Normally solid samples are weighed by difference, that is, the

Figure 2-7

A Mettler H10 single-pan analytical balance.

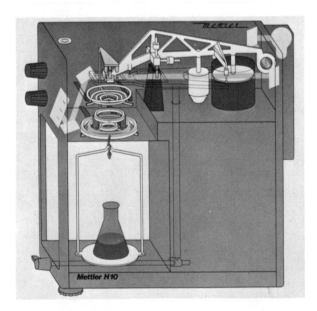

Figure 2-8

Cut-away view of Mettler H10 analytical balance.

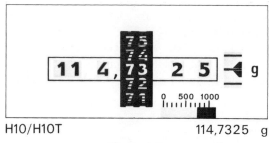

H10/H10T   114,7325 g

Figure 2-9

Optical Vernier Scale for Mettler H10
analytical balance.

complete dry and cool sample contained in a weighing bottle is
weighed. Enough sample is removed with a spatula to comprise the
analytical sample and the weighing bottle containing the remaining
sample is again weighed. The difference in weight is the sample weight.
This process is repeated until the required number of samples has been
removed. This method requires only one more weighing than the number of samples required, whereas weighing by addition requires twice
as many weighings as samples. Weighing by addition into a container
having a large glass surface (beaker or flask) has another source of error due to adsorption of moisture on the surface during the weighing.
When weighing by difference one must, of course, be very sure that all
sample removed from the weighing bottle is transferred into the sample
container. Any sample adhering to the spatula should be brushed into
the container with a camel's hair brush. Any sample spilled necessitates
using a fresh container and obtaining a new bulk weight.

Unless the sample is hygroscopic, the weighing bottle top need not be
weighed with the sample. If the sample is hygroscopic, then the top
must be on the weighing bottle at all times except during the actual
removal of sample. Rubber finger cots or gloves should be worn when
handling the weighing bottle.

Operations of Classical Gravimetric Analysis

Dissolving

The sample is weighed by difference as above into appropriate sized
beakers. The beaker should not have a broken lip and should not be

scratched. Each beaker should contain a glass stirring rod which has been well fire polished. The beaker should be covered with a watch glass whenever possible to prevent dust and other impurities from contaminating the sample. The appropriate solvent should be slowly poured down the side of the beaker to prevent any sample from splashing out. The sample should be dissolved by stirring and heating if necessary.

Precipitation

A transfer pipet, containing the precipitating reagent, is held in one hand against the side of the beaker and the reagent is allowed to flow slowly down the side while the solution is being stirred by a stirring rod held in the other hand.

Filtration

The technique of filtration is independent of the filtering medium used. Only the filtering apparatus itself differs. Figure 2-10 shows filtration through filter paper and Figure 2-11 shows filtration through a sintered glass crucible.

Filtering through Filter Paper

The type of filter paper depends on the particle size of the precipitate. Very fine crystals such as barium sulfate must be filtered on a fine filter paper such as Whatman No. 42. Medium precipitates such as silver chloride can be filtered on medium filter paper such as Whatman No. 40. Whatman No. 41 is similar to No. 40 but it is of a more open texture and, therefore, recommended for coarse precipitates. Gelatinous precipitates, those containing a large amount of water of hydration or adsorbed water, should also be filtered with Whatman No. 41.

If the precipitate is to be ignited and weighed, the filter paper must be ashless, that is, when ignited, a negligible amount of residue remains. Filter paper should not be used with precipitates that are easily reduced because the hot carbon from the paper will act as a reducing agent. When the precipitate is not going to be weighed, for example, as in

Figure 2-10

Filtration with filter paper.

the removal of an interference with subsequent analysis of the filtrate, the use of ashless paper is not necessary.

The following technique will insure efficient filtration: Fold the paper along its diameter (Figure 2-12a). Make a second fold such that the edges do not quite match (about 2 mm apart) (Figure 2-12b). Next form a cone which is three layers thick on one side and one layer thick on the other. Starting at a point about half-way back on the thick side, tear the outer layer down about 1 cm. Continue to tear the paper back towards the open side and fold this back along the single layer portion of the cone (Figure 2-12c). Now place the folded paper in the funnel and wet it with distilled water. The wet filter paper (Figure 2-12d) is pressed against the funnel to make a seal. The stem of the funnel should remain filled with water. This creates a slight suction and increases the rate of filtration. The tip of the funnel should rest with the long edge against the side of the beaker (see Figure 2-10). This increases the suction slightly and prevents splattering of the filtrate. If the filtrate is to be preserved for further analysis, a clean watch glass

Figure 2-11

Filtration with a filter crucible.

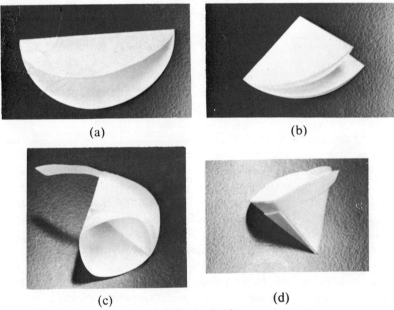

(a)   (b)

(c)   (d)

Figure 2-12

Folding of filter paper.

should be placed over the beaker during filtration to prevent impurities from getting into it.

Filtering through Filter Crucibles

Gooch crucibles, sintered glass filter crucibles, and porcelain filter crucibles are all used for filtering precipitates by suction. A Gooch crucible is the oldest type of filter crucible and is porcelain with holes in the bottom. An asbestos pad or a glass filter disc is placed in the crucible to hold the precipitate. Gooch crucibles are troublesome to prepare and to dry to constant weight. They have been largely replaced with sintered glass filter crucibles or porcelain filter crucibles.

Sintered glass filter crucibles are made of a resistant glass and have a porous sintered ground glass bottom fused onto the crucible. They can only be used with precipitates that are dried at low temperatures. The sintered glass bottom is available in various porosities.

A porcelain filter crucible has a porous unglazed porcelain bottom. They can be heated to a much higher temperature than sintered glass crucibles and, therefore, are useful for a wider range of applications. Porcelain filter crucibles can also be obtained in different porosities.

All three types of filter crucibles (one type, the sintered glass crucible, is shown in Figure 2-13a) are used with a crucible holder (Figure 2-13b) mounted on a filter flask (Figure 2-13c). A safety bottle (Figure 2-13d) is placed between the flask and the aspirator (Figure 2-13e) to prevent tap water from backing up into the filtrate. The aspirator is attached to a service water line. Pressure tubing should be used for all connections.

Transfer of Precipitate

The precipitate should be allowed to settle in the beaker prior to filtration. This allows the bulk of the liquid to be decanted and filtered rapidly before the pores of the filter become clogged. To prevent a loss of small amounts of precipitate, the solution is poured down a glass stirring rod. The final drop of solution clinging to the tip of the rod can be touched off on the inner wall of the beaker or filter. Pouring the decantate down the stirring rod as in Figure 2-14 directs it into the filter without splashing and prevents drippage on the outer lip of the beaker.

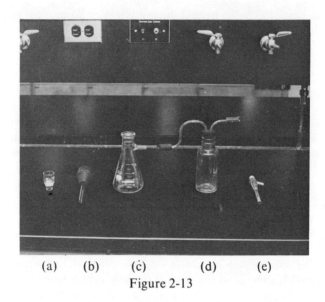

(a)   (b)   (c)   (d)   (e)
Figure 2-13

Suction filtration apparatus: (a) sintered glass
crucible, (b) crucible holder, (c) filter flask,
(d) safety bottle, (e) aspirator.

The precipitate is best washed in the beaker. After the mother liquor
has been removed by decantation, several milliliters of wash solution is
added to the beaker. The precipitate and wash solution are stirred and
then the precipitate is allowed to settle and the wash solution removed
by filtration. This process is repeated two or three times. The precipitate is then transferred to the filter by holding the stirring rod and
beaker in one hand while using the other hand to wash the bulk of the
precipitate down the rod into the filter with a wash bottle. This process
is shown in Figure 2-14.

The last traces of precipitate are removed from the side and bottom
of the beaker by scrubbing them with a moist rubber policeman (Figure 2-15). These traces are also washed down the stirring rod into the
filter.

After all the precipitate is in the filter, it is washed five or six times
with small portions of wash liquid. This is much more efficient than
washing with a single large volume. Each portion should be allowed to
drain before adding the next one. A few drops of the last washing is
collected in a small tube and tested for the presence of the precipitating reagent. If the test is negative, the washing is complete.

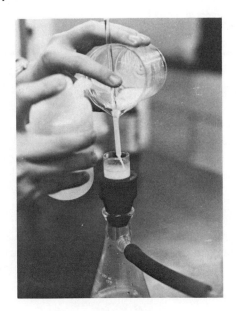

Figure 2-14

Transfer of precipitate into filter.

Figure 2-15

Use of a rubber policeman.

Operations of Volumetric Analysis

Cleaning of Volumetric Glassware

Volumetric glassware must be scrupulously clean. If a film of dirt or grease is present, liquids will not drain uniformly and breaks in the water film or water droplets will form. Under such conditions calibration will be incorrect. Initial cleaning should be done with a warm dilute detergent solution. Cleaning of burets is aided with a buret brush. The wire of the buret brush should be covered with rubber tubing (Figure 2-16) to prevent scratching the inner surface during brushing. Test tube brushes can be used to clean the necks of volumetric flasks.

Often, cleaning with detergent fails to clean glassware satisfactorily. Treatment with a sulfuric acid-dichromate cleaning solution is recommended.* Sulfuric acid-dichromate cleaning solution is extremely corrosive and should only be used with the approval of and under the direct supervision of the laboratory instructor. Hot (60°–100°C) cleaning solution is much more effective than cold cleaning solution but also is more hazardous. Handle with extreme care. Always put cleaning solution in a beaker before heating. Used cleaning solution can be returned, when cooled, to the stock bottle and reused until it becomes green in color.

Burets and small volumetric flasks can be filled with the cleaning solution and drained after treatment. Larger volumetric flasks can be partially filled, tilted and rotated to coat the entire surface, and then drained. Pipets can be cleaned best by repeated filling and draining. The glassware is then rinsed thoroughly with tap water followed by a rinsing with distilled water.

Use of Buret

Once a buret has been cleaned as in the previous section it should be kept clean. A buret that is cleaned correctly at the start of a term

---

*The cleaning solution is prepared as follows: Dissolve 20 g of sodium dichromate in 15 ml of water. Cautiously add 400 ml of concentrated sulfuric acid with stirring. When the heat of reaction has dissipated, the solution can be stored in a wide-mouth, glass-stoppered bottle. This solution is very corrosive so care must be exercised that it not come into contact with skin, clothing, or desk top. Any spills should be immediately flushed with tap water.

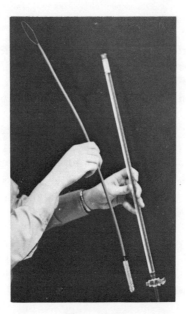

Figure 2-16

Buret brush.

will still be as clean at the end of the term if cared for properly. Whenever a buret is not in use, it should be kept in an upright position filled with distilled water and covered with a small (5 ml) beaker to keep dust particles out.

The distilled water is drained from the buret just prior to filling with titrant. The buret is rinsed with the titrant. Pour 5 ml of titrant into the buret, tilt and rotate the buret to coat the entire surface, and then drain through the stopcock. Repeat this rinsing at least two more times. The liquid coating the walls of the buret now is the same concentration as the titrant and the buret is ready to be filled.

Fill the buret to above the zero mark. After any air bubbles have disappeared, carefully adjust the level of the meniscus exactly to the zero mark. The bottom of the meniscus is where the volume should be read. Care must be taken that your eye is at the same level as the meniscus when the level is read. Otherwise a parallax error will result. Care must also be taken to insure that the tip of the buret is completely filled with solution. Any air bubble left in the tip when the buret is filled to the zero mark will most certainly not be there at the end of the titration. Therefore, an error in the observed volume delivered will occur. When the meniscus has been adjusted carefully to the zero mark, a fraction of a drop may be clinging to the tip. This should be removed by touching it to a glass beaker.

A titration is performed as follows: The sample to be titrated is placed in an Erlenmeyer flask and an indicator added. Insert a clean medicine dropper (Figure 2-17), fill it, and remove a portion of the sample as an "ace-in-the-hole." Rinse the outside of the dropper into the flask. Place the dropper in a safe place since it will be added to the flask later. Hold the flask in your right hand (Figure 2-18) and swirl it with a gentle rotating motion. With practice a vigorous mixing can be accomplished without splashing. The top of the flask remains almost stationary. This motion should be practiced so as not to bump the flask against the buret tip. A magnetic stirrer, Figure 2-19, can also be used to stir the solution. Grasp the stopcock of the buret with the thumb and first two fingers of the left hand as shown in Figure 2-20. This allows maximum control of the stopcock and prevents the accidental loosening of it. Open the stopcock and allow the titrant to enter the flask while it is being swirled. Since a small portion of the sample has been retained, the end point can be approached rapidly. The impending approach of the end point can be seen by the ever widening color change where the titrant hits the solution. As the end point nears, slow down the rate of addition of titrant. It is not necessary, now, to detect the exact point at which the indicator shows a permanent change. However, you should not go too far past the apparent end point. After the apparent end point is reached, add the "ace-in-the-hole" to the flask. Rinse the inside of the dropper into the flask with distilled water. Approach the end point by

Figure 2-17

Removing "ace-in-the-hole."

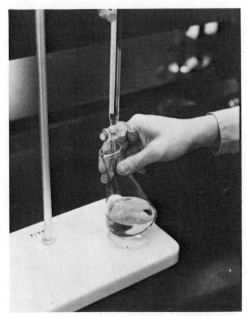

Figure 2-18

Mixing solution during titration.

adding solution drop by drop. Just prior to the end point, again rinse down the sides of the flask. In the vicinity of the end point, add fractional drops. To do this, allow about one-half drop to form on the tip, touch the tip to the side, and rinse down with distilled water. Continue in this manner until the end point is reached as indicated normally by the greatest change in color. If uncertain whether the right color has been reached, record the buret reading and add another one-half drop. If no change occurs the original reading is correct.

An alternative method, used in the vicinity of the end point, is to add fractional drops by a rapid 180° turn of the stopcock. To use this technique hold the barrel of the stopcock steady with the left hand and quickly turn the stopcock with your right hand. This method should only be used with your instructor's approval. Care must be taken not to turn the stopcock too far. Turning 270° and stopping will ruin the titration. Depending on the quickness of the turn, approximately 0.01–0.05 ml of titrant can be added without a drop forming. This eliminates the touching off and washing down of fractional drops. Read the final buret volume as before.

Figure 2-19

Magnetic stirrer.

Figure 2-20

Technique for controlling buret.

If additional titrations are to be made, refill the buret with titrant. Otherwise, drain the buret, rinse well with distilled water, fill above the zero mark with distilled water, and cover with the small beaker until it is again needed. Never leave a solution (especially sodium hydroxide) stand in the buret for long periods of time. It may cause the stopcock to freeze up or may etch the buret.

Occasionally it will be necessary to regrease the stopcock. This occurs when the stopcock ceases to turn easily or when it leaks. To regrease the stopcock remove the stopcock, dry the stopcock and barrel with a paper towel removing the old grease in the process, and remove the old grease from the stopcock hole and buret tip with a fine wire. Then apply a thin layer of grease to the stopcock along the sides 90° from the hole. Place the stopcock into the barrel with the hole aligned. Gently rotate the stopcock back and forth, initially only a few degrees, then steadily increase the rotation until a transparent layer of lubricant covers the whole barrel. The buret is now ready for use.

A problem which sometimes arises during a titration results from the use of too much grease. Some grease becomes lodged in the buret tip and titrant ceases to flow. This does not ruin the determination unless it occurs within approximately 0.5 ml of the end point. (It may not ruin the determination even in this case, but the corrective measure must be done more carefully.) With the stopcock in the open position, hold a lighted match near the buret tip. When the grease softens with heat, the pressure of the liquid will force it out of the tip. The stopcock is turned off and the buret is again ready for use.

Use of Pipet

Two common types of pipets used in quantitative analysis are the volumetric pipet and the measuring or Mohr pipet. These are shown in Figure 2-21. The measuring pipet is calibrated to deliver a variable volume but is not as accurate or precise as a buret. A volumetric pipet is calibrated to deliver a certain volume and is normally as accurate and nearly as precise as a buret when properly used. When exact aliquot portions of a solution are required, a volumetric pipet is used.

When properly cleaned, a pipet will drain uniformly without the formation of droplets. Cleaning instructions were given above. If the pipet is clean and dry, it may be inserted directly into the solution. In this case, draw a few milliliters into the pipet with a rubber suction bulb, tilt and rotate the pipet to wet all the surface, then drain and dis-

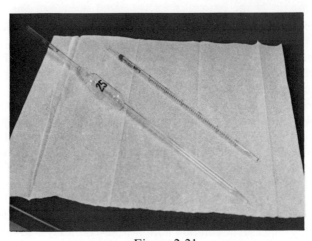

Figure 2-21

Volumetric pipet and measuring pipet.

card the solution. A single rinse suffices when a dry pipet is used. When a wet pipet is used care must be exercised to prevent dilution of the solution. A small amount of the solution is poured into a clean and dry beaker. This solution is used to rinse the pipet as before. At least three rinsings should be made in this case. Now the pipet can be placed in the solution container for filling.

Hold the pipet in one hand, and with a rubber suction bulb in the other hand, fill the pipet to about an inch above the mark. Remove the suction bulb and quickly place your forefinger over the top of the pipet to stop the outflow of solution. Allow the solution to slowly flow from the pipet until the bottom of the meniscus just touches the line by slightly releasing the pressure on your forefinger. Do not remove your finger and attempt to catch the meniscus at the right spot. This could literally take an entire laboratory period to accomplish! Practice is required to become proficient at pipetting. During the entire operation, the hand which holds the pipet is positioned on the narrow portion of the pipet. Never grasp a pipet around the bulb portion because this will cause the solution to warm and expand.

With the meniscus just to the mark and no air bubbles anywhere in the pipet, touch off the droplet clinging to the tip. Tilt the pipet slightly. This causes the solution to flow slightly back from the tip towards the other end. The pipet can now be transported without loss of solution. Use a clean tissue to wipe away any solution clinging to the outer surface of the tip.

Place the tip of the pipet against the inner surface of the vessel to which the solution is being transferred. Remove your forefinger and allow the solution to drain. After the solution has ceased to flow allow the pipet to drain for an additional 20 seconds and touch off any droplet on the tip. Do not blow out the small amount of solution remaining in the tip even though it grows larger with time. The pipet has been calibrated to account for this drainage.

After the pipet has been used, rinse it thoroughly with distilled water. If feasible the pipet should be stored in a cylinder of distilled water protected from dust with a cover. Otherwise, it should be placed in a drawer where it will not be easily scratched or chipped.

Preparation of Standard Solutions

In Chapter 6, we will see that standard solutions can either be prepared directly with a primary standard or can be standardized by titration.

When a solution is to be standardized by titration, special precautions in its preparation are unnecessary. Accurate weight and volume measurements are not required.

When a standard solution is made by dilution of a known weight of primary standard to a known volume, a volumetric flask is used. The volumetric flask is cleaned as above and does not need to be dry. If a predetermined exact concentration is desired, an exact weight must be weighed out. Place a clean and dry watch glass on the balance pan. When using a double-pan balance, a second watch glass of equal or slightly less weight is placed on the right pan as a tare. In either case, record the weight or apparent weight (actual less tare) of the watch glass. Add the exact (within a certain tolerance based on accuracy required) amount of reagent to the watch glass. Transfer the reagent into a small beaker using a stream of distilled water as in Figure 2-22. Dissolve the reagent in distilled water and transfer quantitatively to the volumetric flask. Alternatively the sample can be weighed directly in the beaker, but it is more difficult to adjust the sample to the predetermined weight with a beaker than with a watch glass. Fill the flask about three-quarters full and mix thoroughly. If an appreciable heat of solution occurs, allow the solution to reach room temperature before diluting to the mark with distilled water at room temperature. For highly accurate work, the temperature must be taken when diluted and when used, and a temperature correction applied as described in Chap-

Figure 2-22

Transferring sample to beaker.

ter 6. The solution is properly diluted when the bottom of the meniscus just touches the mark on the flask. Mix the solution thoroughly by repeated alternate inversion and shaking followed by upright shaking.

If an exact concentration of an approximate strength is desired, the reagent is weighed directly in a small beaker by addition. Add the approximate required weight, then calculate the exact concentration from the known weight. The reagent is then dissolved, transferred into the volumetric flask, and diluted as above.

After mixing, transfer the solution into a dry stoppered or capped bottle. Basic solutions should be stored in plastic containers. EDTA solutions should also be stored in plastic containers. Light sensitive solutions should be stored in a dark bottle and kept in a drawer when not in use. If the bottle is wet, use two or three small portions of the standard solution to rinse the bottle. Label the bottle identifying the contents, the concentration, the date of preparation, and the initials of the person preparing the solution. For highly accurate work the temperature at dilution is also shown. Figure 2-23 shows a proper label.

Before use, a standard solution should be well shaken to combine any condensed water at the top of the bottle with the bulk of the solution. Never attempt to fill a buret directly from a standard solution

Figure 2-23

Label for standard solutions.

bottle. It is a good way to "wash" the outside of the buret, its holder, and the desk! Pour the standard solution into a small dry beaker, then fill the buret from the beaker. Between titrations cover the beaker with a watch glass to prevent contamination and evaporation. Discard any unused standard solution, never return it to its original container. There is always a chance of contamination if this were to be done.

General Laboratory Cleanliness

To obtain necessary accuracy and precision in quantitative analysis requires clean apparatus and clean surroundings. When your laboratory period ends, wipe your bench top clean. If any dust accumulates by the next session, wipe the bench before opening your locker. Much laboratory time is lost in cleaning glassware. This is not to imply that it should not be done, rather that the cleaning could be done more efficiently.

Many procedures call for clean and dry beakers. At the start of the term, wash all your beakers in dilute detergent. Rinse with tap water, followed by distilled water. Wipe the excess water from the *outside* with a paper towel. Never wipe the inside as it will leave fibers from the towel. Place the different sized beakers in a nest and invert in your drawer. They will now be clean and dry when needed. After they are used rinse well with small amounts of distilled water, wipe the outside dry, and store as before. It only takes a few seconds at the end of a laboratory period to rinse clean. If left unrinsed for a few days, chemicals may cake on the surface and a more difficult cleaning job results.

Other glassware can be cleaned at the start of the term as above and will be ready to use when needed.

A small wash bottle filled with distilled water, kept on your desk, will save many trips to the distilled water tap.

Laboratory Safety

In discussing laboratory technique, a few words about safety should be mentioned. In any chemical laboratory, potentially hazardous chemicals are used. A mandatory piece of personal safety equipment is safety glasses. Everybody, *including visitors,* in a chemical laboratory should wear some type of eye protection. Ordinary eye glasses (not sun glasses which restrict vision) are suitable in most beginning quantitative analysis laboratories. Tempered safety glasses are much better but the cost of prescription safety glasses precludes their use in beginning courses. Contact lenses are *not* considered eye protection in a laboratory. Certain operations such as mixing of strong acids may require safety glasses with side shields or a full face shield. Consult your instructor if in doubt. Never assume the attitude that since *you* are not working with potentially hazardous chemicals that you may safely remove your eye protection. Usually the person injured in a laboratory accident is not the one who caused the accident but is an innocent bystander!

Most laboratories have the following pieces of standard safety equipment: fume hood, fire extinguisher, fire blanket, safety shower, and eye shower. Learn where these items are located and their proper use!

Proper attire in a laboratory is important. Shorts and short skirts are hazardous to the wearer. The wearing of a laboratory apron is recommended but usually left up to the discretion of the student.

Proper footwear is a necessity. Under no conditions should anyone enter a laboratory without wearing shoes. Small bits of glass and/or chemicals may be on the floor. Likewise open sandals are hazardous. Hot, corrosive liquids are sometimes spilled on the floor. Proper shoes offer much protection against this hazard.

Acid spills can be neutralized with soda-ash and then mopped up with water. Acids spilled on the skin should be immediately flooded with water. Proper attention will prevent chemical burns after exposure has occurred.

First Aid

Never move or lift an accident victim unless he is in danger of further injury. Calm and comfort the victim. If the injuries will permit, summon and wait for a trained person to arrive and administer first aid. If immediate action is needed, one or more of the following may be done while awaiting first aid assistance. Above all remain calm yourself!

1. Acid or alkali burns. Immediately flush the affected area with copious amounts of water. Remove contaminated clothing.
2. Cyanide inhalation. Remove victim to fresh air. Break an amyl nitrite capsule in a handkerchief or gauze pad and hold under the victim's nose, at chin level, for 15–30 seconds, every two minutes. *Do not exceed five applications of amyl nitrite.* Avoid excitement or exercise.
3. Electrical shock. Remove the source of shock. Use a dry towel when moving "live" wires or equipment. If the victim is not breathing, give artificial respiration.
4. Swallowed poison. Dilute strong acids and alkalies by quickly administering large amounts of water. For other poisons, follow instructions on label. Never give liquids to an unconscious person.
5. Bleeding. Hold a clean towel or gauze pad directly on the wound, and apply hand pressure. Do not apply a tourniquet.
6. Not breathing. Start artificial respiration immediately.
7. Fractures. Do not move victim unless *absolutely* necessary.
8. Thermal burns. Do not attempt to remove clothing; immerse burned areas in cold water.

3

Treatment of Analytical Data

Errors

Any measurement made in the laboratory is subject to errors. These errors may be due to the nonhomogeneity of the system being measured, the inaccuracy of the measuring instrument, an operative error, or bias of the observer.

Often in quantitative analysis, steps which precede the measurement step will introduce larger errors than occur in the measurement. The beginning student in quantitative analysis usually cannot do anything to correct nonhomogeneity or inaccuracy of the instrument. The major source of error in beginning courses is due to improper technique of the student.

Classification of Errors

Errors can be divided into two types: (1) determinate and (2) indeterminate errors.

41

Determinate errors, also known as systematic errors, are those whose magnitude can be determined and a proper correction applied. These errors may be constant as in the use of an uncalibrated pipet or an uncalibrated weight. Other determinate errors will vary in magnitude and sign. The expansion and contraction of a standard solution with a change in temperature causes a variable error which can be corrected when the coefficient of expansion for the solution is known.

Some common determinate errors include the following:

1. Instrumental errors
   Uncalibrated glassware
   Uncalibrated weights
2. Operative errors
   Inaccurate standardization of standard solutions
   Failure to apply temperature corrections
   Failure to apply buoyancy corrections
3. Methodic errors
   Solubility of precipitates
   Coprecipitation
   Ash in filter paper
   Side reaction or incomplete reaction
   End point error
4. Impurities in the distilled water and reagents

Indeterminate errors, also known as random errors, are indicated by small fluctuations in the successive values of a measured quantity when these measurements are made by one person under controlled conditions using extreme care.

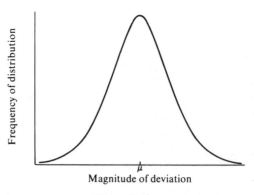

Figure 3-1

The normal distribution curve.

It is commonly said that indeterminate errors are those that the observer cannot prevent nor make allowance for. However, this refers to a certain set of conditions and methods of measurement. By working under different conditions, by employing more sensitive instruments, and by taking certain precautions, it may be possible to reduce the variations in successive measurements.

Because of the nature of their origin, indeterminate errors can be treated by statistical methods. The distribution of indeterminate errors is known to be normal or Gaussian, as shown in Figure 3-1. This curve shows that small indeterminant errors occur more often than large ones and that they are distributed symmetrically about the true value.

Accuracy and Precision

The terms accuracy and precision are often used by laymen synonymously, but it is incorrect to do so. *Accuracy* is the degree of agreement between the measured value and the true value or the accepted true value. *Precision* is defined as the degree of agreement between replicate measurements of the same quantity.

Precision is usually expressed in terms of the deviation of a set of experimental results from the arithmetic mean or average. Precision is a measure of the reproducibility of a result. It implies nothing about their relation to the true value. Although good precision is usually taken as an indication of good accuracy, it is possible to obtain good precision and have poor accuracy. The converse is also true.

Absolute and Relative Error of a Measurement

The accuracy of a measurement may be expressed as either absolute or relative error. The absolute error of a measurement is the difference between the observed value and the true value. An absolute error may be positive or negative and has the units expressed by the two values.

$$\text{absolute error} = X_i - \mu$$

where: X_i = measured value
μ = true value

For example, if the accepted value of chloride in a gravimetric sample is 59.38% and an analyst reports 59.31% the absolute error is

$$59.31\% - 59.38\% = -0.07\%$$

A more common method of expressing error is *relative error*. Relative error is equal to the absolute error divided by the true value and consequently is dimensionless.

$$\text{relative error} = \frac{X_i - \mu}{\mu}$$

Relative error is usually expressed in parts per hundred (%) or in parts per thousand (ppt). For example, the relative error in the above determination is the following:

relative error (%): $\dfrac{59.31 - 59.38}{59.38} \times 100 = -0.12\%$

relative error (ppt): $\dfrac{59.31 - 59.38}{59.38} \times 1000 = -1.2\,\text{ppt}$

The same absolute error, -0.07%, in a sample containing 4.20% chloride (true value) would be the following:

relative error (%): $\dfrac{-0.07}{4.20} \times 100 = -1.7\%$

relative error (ppt): $\dfrac{-0.07}{4.20} \times 1000 = -17\,\text{ppt}$

Significant Figures

The number of significant figures can be defined as the number of digits necessary to express the results of a measurement consistent with the measured precision. As used by most scientists, significant figures are taken to be all the digits that are certain plus one digit which contains some uncertainty. A 50 ml buret is marked at 0.1 ml intervals. A buret reading should be recorded to the nearest 0.01 ml although an error of ±0.01–0.02 may be introduced.

The number of significant figures in a measurement is independent of the placement of the decimal point. For example, 208 millimeters, 20.8 centimeters and 0.208 meters all express the same significance. Changing the units in which a measurement is made does not affect the significance of the number.

The digit 0 can be a significant figure or it may be used merely to locate the decimal point. Zeros placed after the decimal point but before the first nonzero digit are never considered significant. The number 2.03 millimeters can be expressed as 0.00203 meters, but it still contains only three significant figures. Any zeros occurring between nonzero digits are always significant. Zeros occurring after nonzero digits after the decimal point are always significant. The zeros in the numbers 40.00 milliliters, 4.2010 grams, and 2.00 meters are significant. Zeros placed after the last nonzero digit but preceding the decimal point are often ambiguous as to significance. They may or may not be significant depending on the measurement. If a meter stick accurate to the nearest millimeter (1.000 meters) is expressed as 1,000,000,000 millimicrons, only the first three zeros are significant with the remaining six zeros appearing only to locate the decimal point. To avoid this type of ambiguity, large numbers should be written in scientific notation. The above example can be written as 1.000×10^9 millimicrons, clearly showing the correct amount of significance.

In addition and subtraction, absolute uncertainties are involved and only as many decimal places may be retained in the answer as occurs in the entry with the *least* number of decimal places. The molecular weight of lead chromate, $PbCrO_4$, is calculated below.

$$
\begin{array}{lll}
 & Pb & 207.19 \\
 & Cr & 51.996 \\
4 & O & \underline{63.9976} \\
 & & 323.1836 \quad (323.18)
\end{array}
$$

The molecular weight of lead chromate cannot be expressed any more accurately than the nearest 0.01 because the atomic weight of lead is not known any more accurately than to the nearest 0.01. Therefore, the correct molecular weight for lead chromate should be expressed as 323.18.

A standard procedure for rounding off numbers to eliminate nonsignificant figures is employed. When the first digit past the last significant figure is four or less, the insignificant figures are dropped. When the first digit past the last significant figure is six or more, or is a five followed by other nonzero digits, the last significant figure is increased by one. When the insignificant figure is exactly five, the last significant figure is written as an even integer. The following examples serve to illustrate the method, all numbers being significant to the nearest 0.01.

48.2163 round to 48.22
43.3748 round to 43.37
82.1650 round to 82.16
80.1550 round to 80.16
80.1651 round to 80.17

Often in expressing the results of a calculation, one more digit is recorded than is justified by the data. When this is done, the last digit is written as a subscript. The data shown in Figure 2-2 illustrate the use of this method. The normality is expressed as 0.1219_0, 0.1218_6, and 0.1219_3 for the three determinations. When these values are averaged, the extra digit is included and the average is rounded off to the proper number of digits.

In multiplication and division relative uncertainties rather than absolute uncertainties are involved. An improper conception is that the answer should contain the same *number* of significant figures as that included in the data entry comprising the *least* number of significant figures. The correct criteria to apply to an answer is that it should reflect as closely as possible the same relative uncertainty as is found in that data entry which has the greatest uncertainty. In other words, an answer should express about the same uncertainty as the *weakest link in the chain of data* from which the answer was derived. The following example serves as an illustration of the weakest link in the chain of data.

EXAMPLE 3-1: A steel beam is found to be 30.8 feet long and to weigh 634 pounds.

What is the significance of these measurements? The first measurement indicates that the length is closer to 30.8 feet than it is to 30.9 feet or to 30.7 feet. Similarly, the weight is closer to 634 pounds than it is to either 633 pounds or 635 pounds. In other words, the initial measurement is good to 1 part in 308 parts. Note that the location of the decimal point is ignored in determining the significance or uncertainty of a single number. Likewise, the second measurement is significant to one part in 634 parts.

Therefore, the weakest link in the chain of data is the length measurement, which has an uncertainty of 1 part in 308 parts. Because of this weakest link any answer derived as a result of multiplication or division of these two numbers should not reflect an uncertainty different from 1 part in 308 parts. However, in practice it will rarely be possible to express the answer with exactly the same uncertainty as the weakest link.

The purpose of the following calculations is to serve as a guide in determining which of two possible answers to choose, that which expresses greater uncertainty than the weakest link or that which expresses lesser uncertainty than the weakest link. The idea to keep in mind is that the better choice is the answer with an uncertainty which comes closer to approximating the same uncertainty as the weakest link.

Referring to Example 3-1, the following manipulations may be made:

(a) $30.8 \times 634 = \boxed{1.95 \times 10^4}$ or 1.953×10^4

(b) $\dfrac{30.8}{634} = 0.049$ or $\boxed{0.0486}$

(c) $\dfrac{634}{30.8} = \boxed{20.6}$ or 20.58

In each case the correct answer is indicated. The correct answer to (a) is 1.95×10^4 because this represents an uncertainty of 1 part in 195, whereas 1.953×10^4 represents an uncertainty of 1 part in 1953. Obviously 1 part in 195 is the closer approximation to 1 part in 308 parts which was our weakest link. Likewise in (b), 0.0486 represents an uncertainty of 1 part in 486 parts which is a better approximation to 1 part in 308 parts than is represented by 0.049 (1 part in 49 parts). Also in (c), the correct answer is 20.6 which has an uncertainty of 1 part in 206 parts.

In (a), (b), and (c) the correct answer was expressed to three significant figures which was also the number of significant figures in the weakest link. This is not to contradict the statement made earlier about matching the degree of uncertainty or significance of the answer to that of the weakest link rather than matching the number of significant figures. Indeed, in the majority of cases the correct number of significant figures in the answer will equal the number of significant figures in the weakest link. The following two examples will illustrate cases where the correct number of significant figures in the answer does not coincide with the number of significant figures in the weakest link.

EXAMPLE 3-2: $\dfrac{4654 \times 9.02}{382.5} = 110$ or $\boxed{109.7}$

EXAMPLE 3-3: $\dfrac{28.43 \times 58.3}{19.4} = \boxed{85}$ or 85.4

Again the correct answers are indicated.

In Example 3-2 the number of significant figures in the weakest link (9.02) is three, whereas the correct answer has four significant figures.

This is because the answer 109.7 expresses an uncertainty of one part in 1097 whereas the answer 110 expresses an uncertainty of 1 part in 110 parts. The weakest link has an uncertainty of 1 part in 902 parts. Obviously, 1 part in 1097 parts is more representative of the weakest link than 1 part in 110 parts would be.

In Example 3-3 the number of significant figures in the weakest link is three, but the correct answer has only two significant figures. This is because the answer 85 expresses an uncertainty of 1 part in 85 parts while 85.4 expresses an uncertainty of 1 part in 854 parts. The weakest link had an uncertainty of 1 part in 194 parts. To select 85.4 as the correct answer would express an answer having less than one-fourth as much uncertainty (over 4 times the significance) as the weakest link. Selecting 85 as the correct answer would express an answer having 2 times as much uncertainty (one-half as much significance) as the weakest link. The more reasonable choice is to select the answer which has twice the uncertainty of the weakest link rather than that which expresses only one-fourth as much uncertainty.

In the event that the choice of two answers is equivalent (i.e., the uncertainty of one choice is greater than the desired uncertainty by a factor equal to or approximately equal to the factor by which the other answer is less uncertain), it is normally preferable to express the answer as that one which expresses an uncertainty less than the uncertainty of the weakest link.

The discussion on significant figures has been based on the assumption that all data were significant to ±1 in the last digit. If it is known that there exists a greater uncertainty than ±1 in the last digit, appropriate calculations can be applied to the uncertainties.

Now that we have learned how to express our results with the appropriate degree of uncertainty we can turn to the evaluation of the results.

Statistical Treatment of Data

Mean

The central tendency of a group of data is that value about which individual entries tend to cluster. The arithmetic mean or average is taken as the best measure of central tendency. It is calculated as follows:

$$\overline{X} = \frac{X_1 + X_2 + \cdots + X_n}{n} = \frac{1}{n} \sum_{i=1}^{n} X_i$$

where: $\overline{X}$ = mean
X_1, X_2, etc. = value of individual results
n = total number of results

It can be shown that the mean of n results is $\sqrt{n}$ times as reliable as any of the individual results. This means that there is a diminishing return from accumulating more and more replicate measurements. The mean of 4 results is twice as reliable as one result; the mean of 9 results is three times as reliable; the mean of 16 results is four times as reliable, etc.

EXAMPLE 3-4: Calculate the mean of the following results: 21.32% Cl, 21.46% Cl, 21.38% Cl, 21.51% Cl, 21.29% Cl.

$$\overline{X} = \frac{21.32 + 21.46 + 21.38 + 21.51 + 21.29}{5} = 21.39\%$$

Range

The simplest measure of variability about the central tendency is the range, which is the difference between the highest and lowest values. The range has been recommended as a better estimate of precision than average or standard deviations when n is from 3 to 10.* However, in general, the range is an inefficient measure of variability because a single result with a large error exerts its full impact upon the range; but its effect is diluted by the other results when the average deviation or standard deviation is used.

The range for the determinations in Example 3-4 is

$$21.51\% - 21.29\% = 0.22\%$$

---

*R. B. Dean and W. J. Dixon, *Anal. Chem.*, **23**, 636 (1951).

Average Deviation

The average deviation from the mean $\bar{d}$ is often used as a measure of variability but is not statistically significant. Average deviation is easy to calculate, being the average of the absolute values of all the deviations.

$$\bar{d} = \frac{1}{n} \sum_{i=1}^{i=n} |X_i - \bar{X}|$$

The calculation of the average deviation for the data in Example 3-4 follows with all entries as percentages:

X_i	$|X_i - \bar{X}|$
21.32	0.07
21.46	0.07
21.38	0.01
21.51	0.12
21.29	0.10

$$\bar{d} = \frac{0.07 + 0.07 + 0.01 + 0.12 + 0.10}{5} = 0.07_4$$

The average deviation can be expressed on a relative basis as well as an absolute basis.

$$\text{relative average deviation} = \frac{\bar{d}}{\bar{X}} \times 1000$$

$$= \frac{0.07_4}{21.39} \times 1000 = 3.5 \text{ ppt}$$

Standard Deviation

The standard deviation s is a much more statistically significant measure of variability than is average deviation. Standard deviation is the root-mean-square deviation of values from their average. Standard deviation is calculated as follows:

$$s = \sqrt{\frac{\Sigma(X_i - \bar{X})^2}{n - 1}}$$

The calculation of the standard deviation for the data in Example 3-4 follows, again with all entries as percentages:

X_i	$|X_i - \overline{X}|$	$(X_i - \overline{X})^2$
21.32	0.07	0.0049
21.46	0.07	0.0049
21.38	0.01	0.0001
21.51	0.12	0.0144
21.29	0.10	0.0100

$$s^2 = \frac{0.0049 + 0.0049 + 0.0001 + 0.0144 + 0.0100}{4}$$

$$= 0.0086$$

$$s = \sqrt{.0086} = 0.09_3$$

When the standard deviation is expressed as a percentage of the mean, it is called the *coefficient of variance, v.*

$$v = \frac{s}{\overline{X}} \times 100$$

$$= \frac{0.09_3}{21.39} \times 100 = 0.43\%$$

The standard deviation can also be expressed on a relative basis.

$$\text{relative standard deviation} = \frac{s}{\overline{X}} \times 1000$$

$$= \frac{0.09_3}{21.39} \times 1000 = 4.3 \text{ ppt}$$

The square of the standard deviation s^2 is known as the *variance.*

Confidence Limits

When an infinite number of measurements are made where the only errors are random errors, the mean value $\overline{X}$ will equal the true value μ. This infinite number of measurements is known as a population. The standard deviation for such a population is known as σ and the per-

Treatment of Analytical Data

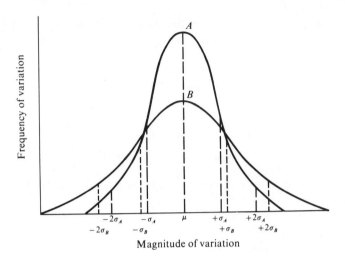

Magnitude of variation

Figure 3-2
Two populations having the same central tendency but
different variances.

centage of results lying within multiples of σ are accurately known.
Figure 3-2 shows two normal error curves or normal distribution curves
having the same central tendency μ but different variabilities σ_A and σ_B.
For both curves 68.26% of all results lie within plus or minus one
standard deviation $(\pm 1\sigma)$, 95.46% of all results lie within $\pm 2\sigma$, and
99.74% of all results lie within $\pm 3\sigma$. Table 3-1 shows the percentage of
results with other factors of σ. Using this distribution of an infinite
population one can predict precisely the odds of drawing from this
population an observation lying inside or outside certain limits.

Unfortunately, an analyst does not have the time nor resources to
perform an infinite number of determinations. Consequently, the mean
and standard deviation are not well defined. If $\overline{X}$ and s are taken as
estimates of μ and σ, confidence limits calculated from Table 3-1 will
give too favorable results. The probability of a deviation being greater
than a specified value will be underestimated because the estimated
value of σ will be too small. An English chemist W. S. Gosset, writing
under the pen name of Student, studied the problem of confidence limits based upon a finite sample drawn from an unknown population and
published a solution which is now widely accepted.* The effect of the
number of observations on the confidence limits is calculated with the

---

*Student, *Biometrika*, **6**, 1 (1908).

Table 3-1

Normal Distribution of Deviations
in an Infinite Population

Limits for deviation $\dfrac{X_i - \mu}{\sigma}$	Percent of deviations within these limits
0.10	7.96
0.25	19.74
0.50	38.30
0.75	54.68
1.00	68.26
1.25	78.88
1.50	86.64
1.75	91.98
2.00	95.44
2.50	98.92
3.00	99.74
4.00	99.994
5.00	99.99994

aid of t (often called Student's t) which is defined as

$$\pm t = (\overline{X} - \mu)\,\frac{\sqrt{n}}{s} \tag{3-1}$$

where n is the number of observations and all other terms are as previously defined. Several values of t are given in Table 3.2.

To obtain confidence limits, the standard deviation s is calculated and a value for t is found from Table 3-2 based on the probability level and the number of measurements made on the sample.* Statisticians prefer to use $n - 1$, the degrees of freedom, for determining confidence limits.

Rearrangement of Equation (3-1) gives

$$\mu = \overline{X} \pm \frac{ts}{\sqrt{n}} \tag{3-2}$$

---

*Many tables list the probability levels in decimal form called confidence levels or as 1.000 − confidence level called risk factors. For example the right-hand column in Table 3-2 may be headed by 0.99 and be called a confidence level or may be headed by 0.01 and be called a risk factor.

Table 3-2

Values of t for Calculating Confidence Limits

Number of measurements, n	Degrees of freedom, $n - 1$	Probability level			
		50%	90%	95%	99%
2	1	1.000	6.314	12.71	63.66
3	2	0.8165	2.920	4.303	9.925
4	3	0.7649	2.353	3.183	5.841
5	4	0.7407	2.132	2.776	4.604
6	5	0.7267	2.015	2.571	4.032
7	6	0.7176	1.943	2.447	3.707
8	7	0.7111	1.895	2.365	3.500
9	8	0.7064	1.860	2.306	3.355
10	9	0.7027	1.833	2.262	3.250
15	14	0.6924	1.761	2.145	2.510
16	15	0.6912	1.753	2.132	2.490
20	19	0.6876	1.729	2.093	2.861
21	20	0.6870	1.725	2.086	2.845
30	29	0.6830	1.699	2.042	2.750
31	30	0.6828	1.697	2.021	2.705
∞	∞	0.6745	1.645	1.960	2.576

If Equation (3-2) is applied to Example 3-4, the following confidence limits are obtained:

$$50\%: \quad 21.39 \pm \frac{(0.7407)(0.093)}{\sqrt{5}} = 21.39\% \pm 0.03_1\%$$

$$90\%: \quad 21.39 \pm \frac{(2.132)(0.093)}{\sqrt{5}} = 21.39\% \pm 0.08_9\%$$

$$95\%: \quad 21.39 \pm \frac{(2.776)(0.093)}{\sqrt{5}} = 21.39\% \pm 0.11_5\%$$

$$99\%: \quad 21.39 \pm \frac{(4.604)(0.093)}{\sqrt{5}} = 21.39\% \pm 0.19_1\%$$

The correct interpretation of these limits shows that there is a 50% probability that the range 21.36% to 21.42% includes the population mean μ; there is a 90% probability that the range 21.30% to 21.48%

includes μ; there is a 95% probability that the range 21.27% to 21.51% includes μ; there is a 99% probability that the range 21.20% to 21.58% includes μ.

Rejection of Dubious Results

When a series of replicate analyses is made, the analyst is sometimes faced with the problem as to whether a particular result should be included in the statistical treatment of data or be rejected. It goes without comment that when an error was known to be made (spillage, contamination, etc.) the result should automatically be rejected. The problem which arises is whether to discard as questionable a result when no known cause for its deviation is known.

If the number of replicate values is large, a single errant value will have little effect upon the mean, therefore, to reject or not to reject is of no real consequence. However, in the real world of analytical chemistry, the number of replicate values is usually quite small. The erroneous result exerts a more significant effect upon the mean than is warranted, but there are not sufficient data to allow a meaningful analysis of the result.

Rules for the rejection of results have been developed based upon average deviation, standard deviation, and range. Other criteria are also sometimes used.

To reject a result using average deviation, calculate the mean $\overline{X}$, omitting the questionable result. If the deviation of the questionable result from $\overline{X}$ is greater than four times the average deviation excluding that result, then the result should be rejected. A serious limitation to this rule is that it should only be applied to a set of results containing at least four values.

If $| X_? - \overline{X} | > 4\overline{d}$, reject $X_?$.

To reject a result using standard deviation s, calculate s omitting the questionable result. If the deviation of the questionable result from $\overline{X}$ is greater than three times the standard deviation excluding that result, then the result should be rejected. This rejection is statistically based upon the normal distribution curve and consequently does not hold strictly for a sample, particularly when n is less than ten.

If $| X_? - \overline{X} | > 3s$, reject $X_?$.

A third method for rejection of questionable results is the *Q-test*. The *Q*-test is statistically valid and easy to use. However, when applied to from three to five results, the *Q*-test only allows rejection of grossly erroneous results but does not allow the rejection of suspicious but less divergent results.

The *Q*-test is applied as follows:

1. Calculate the range of results.
2. Calculate the difference between the suspected result and its nearest neighbor.
3. Divide the difference in (2) by the range in (1) to obtain a rejection quotient *Q*.
4. Consult a table of *Q* values, see Table 3-3. If the computed value for *Q* is greater than the value in the table, the result can be rejected with a 90% confidence. Values for *Q* are available at other confidence levels.

Table 3-3

Values of Rejection Quotient *Q*

Number of observations	$Q0.90$
3	0.94
4	0.76
5	0.64
6	0.56
7	0.51
8	0.47
9	0.44
10	0.41

Source: R. B. Dean and W. J. Dixon, *Anal. Chem.*, **23,** 636 (1951). Reprinted by permission of the American Chemical Society.

The three criteria for rejection of results are illustrated in Example 3-5.

EXAMPLE 3-5: The following results were obtained for the normality of a sodium hydroxide solution: 0.1216, 0.1235, 0.1219, 0.1216, 0.1215, 0.1220, 0.1210. Apply the average deviation, standard deviation, and *Q*-tests to these data to determine if 0.1210 and/or 0.1235 can be rejected.

Average deviation test:
Is 0.1235 valid?

$$\overline{X} = \frac{0.1216 + 0.1219 + 0.1216 + 0.1215 + 0.1220 + 0.1210}{6} = 0.1216$$

$$\overline{d} = \frac{0.0000 + 0.0003 + 0.0000 + 0.0001 + 0.0004 + 0.0006}{6} = 0.0002_3$$

The questionable deviation is $0.1235 - 0.1216 = 0.0019$ which is more than 4 times $\overline{d}$, therefore, the 0.1235 value can be rejected.

Is 0.1210 valid?

$$\overline{X} = \frac{0.1216 + 0.1219 + 0.1216 + 0.1215 + 0.1220}{5} = 0.1217$$

$$\overline{d} = \frac{0.0001 + 0.0002 + 0.0001 + 0.0002 + 0.0003}{5} = 0.0001_8$$

The questionable deviation is $0.1217 - 0.1210 = 0.0007$ which is slightly less than 4 times $\overline{d}$, therefore, the 0.1210 value cannot be rejected.

Standard deviation test:
Is 0.1235 valid?

$$s = \sqrt{\frac{0 + 9 \times 10^{-8} + 0 + 1 \times 10^{-8} + 16 \times 10^{-8} + 36 \times 10^{-8}}{5}}$$
$$= 0.0003_5$$

The questionable deviation is 0.0019 which is more than 3 times s, therefore, the 0.1235 value can be rejected.

Is 0.1210 valid?

$$s = \sqrt{\frac{1 \times 10^{-8} + 4 \times 10^{-8} + 1 \times 10^{-8} + 4 \times 10^{-8} + 9 \times 10^{-8}}{4}}$$
$$= 0.0002_2$$

The questionable deviation is 0.0007 which is greater than 3 times s, therefore, the 0.1210 value can be rejected.

Q-test:
Is 0.1235 valid?

$$\text{range} = 0.1235 - 0.1210 = 0.0025$$

The difference between 0.1235 and its nearest neighbor is 0.0015. The rejection quotient is $0.0015/0.0025$ or 0.60 which is larger than that required for rejection when $n = 7$ (0.51). Therefore, the value 0.1235 can be discarded.

Is 0.1210 valid?
The range is now recalculated for the six remaining values.

$$\text{range} = 0.1220 - 0.1210 = 0.0010$$

The difference between 0.1210 and its nearest neighbor is 0.0005. The rejection quotient is $0.0005/0.0010$ or 0.50 which is smaller than that required for rejection when $n = 6$ (0.56). Therefore, the value 0.1210 cannot be discarded.

   The high value (0.1235) can be discarded by any of the three tests and indeed must be due to some unknown determinate error. The low value (0.1210) can be discarded by the standard deviation test (bear in mind that this test does not hold strictly for sets of less than 10 values). The average deviation test does not allow the rejection of the low value although it is indeed a borderline case. The Q-test definitely does not permit the rejection of the low value. Obviously the rejection tests do not always yield the same conclusion. In cases where a highly accurate value is needed, the answer to whether a borderline value should or should not be discarded is to obtain more results.

Questions

1. Define the following terms: determinate and indeterminate errors, accuracy, precision, significant figures, mean, range, average deviation, standard deviation, coefficient of variance, variance, confidence limits, Q-test.
2. The magnitude of an error can be expressed in two ways. Name the two ways and explain their difference.
3. An error can be classified as belonging to one of two major types. Name the two types, explain their difference, and list a few examples of each type.

4. What is the difference between the absolute uncertainty and relative un-certainty of a measurement or mean of measurements?

5. Explain when a zero in a measurement is significant and when it is not significant.

6. Write the number 10,000 so as to indicate a relative uncertainty of 1 part per thousand.

7. Describe exactly what is meant by confidence limits.

8. Describe three different criteria which can be used to reject a dubious result. Indicate what limitations are placed on each method.

9. Is it possible to have precision without accuracy? Is it possible to have accuracy without precision? Explain.

Problems

1. Express each of the following numbers with the proper number of digits to indicate a relative uncertainty in parts per thousand closest to that given in parentheses.
   (a) 0.07826 ($\pm$1.4 ppt)   (b) 42.1350 ($\pm$0.5 ppt)
   (c) 4.2146 ($\pm$3.9 ppt)   (d) 28762 ($\pm$5 ppt)

2. Express the result of each of the following calculations to the proper number of significant figures.
   (a) $6.1275 + 1.82 + 0.0637 = 8.0112$
   (b) $4.2831 - 0.024 = 4.2591$
   (c) $3.16 - 1.2431 = 1.9169$
   (d) $16.1 + 0.0243 - 14.3241 = 1.8002$

3. Express the result of each of the following calculations to the proper number of significant figures.
   (a) $42.36 \times 5.98 = 253.3128$ (b) $0.0086 \times 2.125 = 0.018275$
   (c) $0.9824 \div 3.12 = 0.31487$ (d) $47.98 \div 1.2436 = 3.85815$

4. Express the result of each of the following calculations to the proper num-ber of significant figures.

   (a) $\dfrac{4.223 \times 0.00463}{48.2} = 0.00040565$

   (b) $\dfrac{7.001 \times 14.98}{1.23} = 85.264$

   (c) $\dfrac{0.2786 \times 47.64}{94.2} = 0.140897$

   (d) $\dfrac{3.65}{0.01243 \times 14.6} = 20.11263$

5. The following data were obtained for the percentage of chloride in a sample: 53.23, 53.16, 53.31, 53.24, 53.19. Calculate (a) the mean, (b) the range, (c) the absolute standard deviation, and (d) the standard deviation in parts per thousand.

6. Ten students measured the temperature of a solution reporting the following values: 25.4, 25.6, 25.5, 25.5, 25.7, 25.5, 25.3, 25.6, 25.5, and 25.6°C. Calculate the mean and standard deviation of the mean of the values.

7. The following results were made of the atomic weight of sulfur: 32.0628, 32.0653, 32.0636, 32.0660, 32.0641, 32.0638, 32.0631, 32.0634. Calculate (a) the mean, (b) the standard deviation of the mean, and (c) the atomic weight of sulfur at the 99% probability level.

8. An analyst obtained the following values for the normality of a silver nitrate solution: 0.0990, 0.0994, 0.0992, and 0.0991. What is the range for the 95% confidence limits?

9. The following readings were taken for the absorbance of a colored species: 0.222, 0.218, 0.215, 0.222, 0.230, 0.224, 0.223. Calculate the confidence limits at the 90% and 95% probability level.

10. The following data were obtained for the normality of a sodium hydroxide solution: 0.1122, 0.1126, 0.1138, 0.1124, 0.1122, 0.1118. Determine whether any of the results can be rejected on statistical grounds. Calculate the confidence limits at the 95% level.

11. A steel sample was analyzed for manganese by a spectrophotometric method. The following results were obtained: 0.82, 0.85, 0.83, 0.82. If an additional determination were made, calculate how high (minimum) it would have to be before it could justifiably be discarded by the Q-test.

12. Apply the Q-test to the following sets of data to see if rejection of a dubious result is justifiable.
    (a) 41.32, 40.56, 41.53, 41.06
    (b) 84.63, 84.55, 84.87, 84.68

13. Apply the Q-test to the following data for the percentage iron to see if any dubious results should be discarded: 5.63, 5.66, 5.52, 5.60, 5.82, 5.70, 5.64, 5.59.

14. Apply the average deviation and standard deviation tests to problem 13. What conclusions can be drawn from the results?

15. The following percentages of nitrogen in a sample were determined by the Kjeldahl method: 7.32%, 7.46%, 7.50%, and 7.38%.
    (a) Calculate the absolute and relative (parts per thousand) standard deviation of the mean.
    (b) The accepted value for nitrogen in the sample was 7.45%. Calculate the absolute and relative accuracy of the mean.

16. The following results for the percentage of chloride in pure potassium chloride were determined by a technician: 47.62, 47.68, and 47.64.
    (a) Calculate the absolute and relative standard deviation of the mean.
    (b) Calculate the absolute and relative accuracy of the mean.

4

Gravimetric Analysis

Gravimetric analysis is the branch of quantitative analysis in which the percent of a substance in a mixture is determined by its weight. In a classical gravimetric analysis, the substance to be determined is precipitated and the precipitate is isolated and weighed. Gravimetric procedures although time consuming are among the most accurate methods of analysis. In order to perform a successful gravimetric determination, three criteria must be met.

1. The precipitation must be quantitative. To be quantitative at least 99.9% of the substituent being determined must be precipitated. Since an excess of the precipitating reagent is always used in gravimetric analysis, a small excess amount of common ion is always present. This reduces the solubility of the precipitate to such an extent that somewhat moderately soluble precipitates can sometimes be utilized in a gravimetric procedure. It should be noted that in most gravimetric procedures much more than 99.9% of the desired substituent is precipitated.

2. The precipitate must be of a known composition at the time the final weighing is performed. Any impurities in the precipitate not

63

removed during the washing of the precipitate must be volatile enough to be removed during the drying of the precipitate.
3. The precipitate must be in a form which is suitable for filtration.

The steps involved in a typical classical gravimetric analysis are as follows:

1. sampling
2. drying
3. weighing
4. dissolving
5. preparation of solution
6. precipitation
7. digestion
8. filtration
9. washing
10. drying or igniting
11. weighing
12. calculation of results

Sampling

In a beginning course in quantitative analysis, samples are usually in the form suitable for drying, weighing, and dissolution. However, the most critical step in any analytical procedure is the first step, the procurement of a representative sample for analysis. Obviously, no analysis can be any better than the sample used. Tedious and accurate work will be wasted on a sample whose composition is not identical to that of the bulk of the sample.

If the sample is a homogeneous material such as a true solution, any portion of it may be taken for analysis. Many natural materials such as ores, and some manufactured materials such as alloys, are heterogeneous and a large gross or composite sample must be taken. This gross sample which may consist of several kilograms is taken by combining numerous small samples taken at random from various portions of the bulk material. Precise procedures are specified by various professional societies for sampling certain types of materials. After the gross or composite sample has been collected, it is crushed, pulverized, or ground into a uniform particle size. The sample is then subdivided by a process known as quartering into a laboratory sample which is then used for analysis. The technique of quartering involves mixing the sample and pouring into a conical pile. The cone is flattened and divided into quarters. Two diagonally opposite quarters are retained and the other two discarded. The retained quarters are combined and mixed into a new pile which is quartered as before. This process is repeated until the portion remaining is of satisfactory size for a laboratory sample.

Drying the Sample

Many samples, particularly finely divided samples, adsorb moisture on their surface. The amount of moisture adsorbed depends not only on the nature of the sample but also on the atmospheric humidity. A change in the percentage of water in the sample will necessarily cause a corresponding change in the percentages of all other constituents as well. In order for the analysis of a sample to be meaningful, the analysis should be run on a sample in which the adsorbed water has been removed or in which the amount of adsorbed water is known. Drying of most samples is done in an open weighing bottle in an oven at 100° to 110°C for one or two hours. Many samples can be dried for much longer periods of time without danger, but the effect of prolonged heating should be checked out on unknown samples.

Some samples such as alloys and metals do not adsorb moisture appreciably. These samples can be analyzed on an "as received" or air dry basis. Some samples including many organic compounds are partially decomposed even at 100°C and should be dried in a vacuum desiccator or a desiccator at room temperature using a drying agent. Other samples may undergo oxidation when heated or when dried by a desiccant even at room temperature and must not be dried but run on an as received basis. An example of this type is hydrated ferrous ammonium sulfate.

Once the sample has been dried in an oven, it must be placed in a desiccator to cool. After cooling for at least 30 minutes the sample may be weighed. If the dried sample is to be kept for a period of time before weighing, the top should be placed on the weighing bottle.

There are occasions when a sample with a variable water content is analyzed on an as received basis, a separate portion is analyzed, and the desired constituent on the dry basis is calculated from the data. The following examples illustrate the technique:

EXAMPLE 4-1: A sample of soda ash weighing 1.064 g lost 0.0465 g on drying to constant weight at 110°C. A separate undried sample was found to contain 92.4% Na_2CO_3. Calculate the percentage of Na_2CO_3 in the dry sample.

To solve any "wet-dry" type problem it is necessary to remember the fact that the removal of a volatile constituent does not affect the weight of any of the nonvolatile constituents. It only affects the percentage of the nonvolatile constituents in the sample because it alters the sample weight.

It will simplify the solution of any wet-dry problem to assume 100 g of sample on an as received basis. Although any sample weight would do as well, 100 g allows percentages to be converted directly into sample weights.

To solve Example 4-1, first determine the percentage of water.

$$\%H_2O = \frac{0.0465 \text{ g} \times 100}{1.0642 \text{ g}} = 4.37\%$$

In 100 g of as received sample there is 92.4 g of Na_2CO_3, 4.37 g H_2O, and a total of 95.63 g of nonvolatile material. Therefore, the percentage of Na_2CO_3 on the dry basis is given by

$$\%Na_2CO_3 \text{ (dry)} = \frac{92.4 \text{ g} \times 100}{95.63 \text{ g}} = 96.6\%$$

The reverse calculation is also sometimes involved.

EXAMPLE 4-2:   A sample of soft coal when dried to constant weight in an oven lost 6.94% of its weight. The percentage of sulfur in the dry sample was 0.32%. Calculate the percentage of sulfur in the as received sample.

The weight of water in 100 g of the as received sample is 6.94 g, therefore, the weight of nonvolatile is 93.06 g. Of the nonvolatile, 0.32% is sulfur.

$$\text{wt of sulfur} = 93.06 \text{ g} \times 0.0032 = 0.29 \text{ g}$$

This is not only the weight of sulfur in the dried sample but also the weight of sulfur in the as received sample. Since the as received sample was 100 g the weight of sulfur in grams equals the percentage of sulfur.

$$\%S = \frac{0.29 \text{ g} \times 100}{100 \text{ g}} = 0.29\%$$

A third variation of the wet-dry problem arises when only a portion of the moisture is removed. This occurs in air drying and certain other applications.

EXAMPLE 4-3: A sample of cow's milk contains 87.82% water and 4.02% butterfat. The milk is evaporated until the water content is reduced to 24.30%. Calculate the percentage of butterfat in the evaporated milk sample.

Again the solution to this problem is simplified by assuming a 100 g sample. Of the 100 g, 4.02 g is butterfat and a total of 12.18 g is nonvolatile components. After evaporation there is still 12.18 g of nonvolatile components but now the 12.18 g is 75.70% of the evaporated milk. The total weight of evaporated milk is given as follows:

$$\text{wt of milk} = \frac{12.18 \text{ g} \times 100}{75.70} = 16.09 \text{ g}$$

The percentage of butterfat is, therefore,

$$\frac{4.02 \text{ g} \times 100}{16.09 \text{ g}} = 24.98\%$$

The important point in Example 4-3 is that the 24.30% water and the 4.02% butterfat are fractions of different sample weights.

By now one might say "very nice calculations . . . but do they serve any useful purpose?" Yes, there are definite situations in which this type of calculation is useful. First, there is the situation in which as rapid a determination as possible is desired. If the sample were to be analyzed in the conventional manner a delay of 1.5 to 2.5 hours would be necessary before weighing the sample. In volumetric analysis 1.5 to 2.5 hours is usually sufficient to complete a determination. Therefore, if separate samples are weighed out, one for a moisture determination and the other for analysis, the complete determination can be completed in 1.5 to 2.5 hours, cutting the elapsed analysis time by one half. In routine control work, for example analysis on a batch operation, this saving of time could be very significant.

Secondly, some samples when completely dried are so hygroscopic that they make weighing of samples for analysis very difficult. Therefore, separate determinations are preferred.

A third application is in the comparison of assay results between two labs. For example, the seller of a load of ore might report 2.12% H_2O and 62.15% Fe_2O_3. The buyer's laboratory upon receipt might report 1.84% H_2O and 62.35% Fe_2O_3. Calculation of both reports on a dry basis shows that the analyses by the two laboratories agree within 0.02% Fe_2O_3.

In a course in quantitative analysis if there is any doubt as to whether or not to dry a sample prior to analysis, the instructor in charge should be consulted for proper instructions.

Weighing the Sample

In a gravimetric analysis a sample weight should be selected such that the quantity of precipitate will be great enough to provide adequate accuracy in weighing and not too much as to cause problems with the washing and filtration.

The sample should always be handled for weighing in a clean and previously dried glass stoppered weighing bottle.

Dissolving the Sample

Many samples encountered in a beginning course in quantitative analysis are inorganic in nature and can be readily dissolved in distilled water. Others can be dissolved in dilute hydrochloric acid solution. Metals and alloys are normally brought into solution with nitric acid, a strong oxidizing acid.

Perchloric acid, when hot, is an extremely strong oxidizing agent and can be used to decompose difficultly oxidizable material. Extreme care must be exercised with hot perchloric acid because it will react explosively with easily oxidizable material, especially organic matter.

Refractory materials are not completely soluble in acids and must be brought into solution by fusion at a high temperature with compounds known as fluxes. Fluxes owe their solvent properties to the high temperature used, the high concentration of reactive agent, and the formation of soluble reaction products. Sodium carbonate, a basic flux, is commonly used to dissolve silica and other acidic oxides. Potassium pyrosulfate, an acidic flux, is used to dissolve basic substances such as ferric oxide. Sodium peroxide, alone or mixed with sodium carbonate or sodium hydroxide, forms an effective alkaline oxidizing flux used when sodium carbonate alone is ineffective.

In general, methods of analysis include a description of the procedure for dissolving the sample.* Organic materials such as plant and animal tissue and biological fluids are usually decomposed by wet digestion or by dry ashing. Wet digestion involves boiling the sample in a mixture of nitric and sulfuric acid to oxidize the organic matter to carbon dioxide, water, and other volatile products leaving salts or acids of the inorganic constituents. Dry ashing involves the burning off of the organic matter using atmospheric oxygen as the oxidant, leaving an inorganic residue. Dry ashing may or may not utilize an oxidizing aid and occurs at $400°-700°C$ in a muffle furnace.

Preparation of Solution for Precipitation

Now that the sample has been brought into solution, it is necessary to adjust the conditions so that the criteria for a successful gravimetric determination will be met. Thus, it may be necessary to adjust the pH of the solution to allow quantitative precipitation. For example, calcium oxalate is insoluble in basic or neutral solution, whereas at low pH, competition for the oxalate ions occurs between the calcium ion and the hydrogen ion. At a sufficiently low pH, all the oxalate ion is converted into oxalic acid, a weak acid, and the calcium oxalate is completely soluble.

A preliminary separation may be necessary to eliminate interfering ions although preferably the precipitation step itself is selective enough so that a pure precipitate of a known composition will result.

Selection of the proper conditions for precipitation will also aid in satisfying the third general criterion, that the precipitate be in a form suitable for filtration.

Precipitation

The size and form of particles forming a precipitate depend upon the conditions under which the precipitate was formed, the characteristics of the particular substance, and its treatment after precipitation.

A precipitate separates from a supersaturated solution by the forma-

---

*A list of common solvents and fluxes used in inorganic analysis is given by I.M. Kolthoff et al., *Quantitative Chemical Analysis* 4th ed. (New York, The Macmillan Company, 1969), p. 523.

tion of small clusters of ions known as nuclei. Crystal growth occurs through deposition of material from the solution upon the nuclei. In general, the particle size of a precipitate depends on two factors, the rate of formation of the nuclei (called nucleation) and the rate of growth by the nuclei. When conditions are such that the rate of nucleation is very large compared to the rate of growth by the nuclei, the precipitate will be composed of a large number of very small particles which may or may not be filterable. This situation exists when the precipitation occurs from a concentrated or highly supersaturated solution. When the conditions are altered so that supersaturation is low, the number of nuclei which form is relatively small and the particle size is larger.

A precipitate must consist of particles large enough that none can pass through the filtering medium. In addition, the larger the crystals the smaller the total surface area per unit weight of precipitate and, therefore, the less impurities present. Considering these facts it is imperative to keep the amount of supersaturation as low as possible.

Von Weimarn* discovered that the particle size of precipitates is *inversely proportional* to the relative supersaturation of the solution during the precipitation:

$$\text{relative supersaturation} = \frac{Q - S}{S}$$

where Q is the molar concentration of the mixed reagents *before* precipitation occurs and S is the molar solubility of the precipitate at equilibrium. $Q - S$ gives the *degree of supersaturation*. This ratio is also known as the *von Weimarn ratio*.

The larger the von Weimarn ratio, the higher the rate of nucleation, and thus the smaller the particle size. In fact, if $Q - S$ becomes too great, nucleation may predominate over growth and a colloidal precipitate will occur.

To obtain the largest particle size, conditions should be adjusted so that Q is as low as possible and S is as large as possible.

The following techniques are commonly used:

1. *Precipitation from a dilute solution.* This keeps Q low.
2. *Slow addition of precipitating reagent with thorough stirring.* The precipitating reagent should be added slowly from a pipet with one hand while constantly stirring with the other hand. This prevents the formation of a large local excess concentration of reagent. It

---

*P.P. von Weimarn, *Chem. Revs.*, **2**, 217 (1925).

is impossible to attain equal concentration of reagent throughout the solution, but this technique minimizes the effect and keeps Q as low as possible.

3. *Addition of reagents which increase the solubility of the precipitate.* Most precipitates are more soluble in an acid solution than in a neutral solution. When S is larger, the rate of nucleation is slower, and the precipitate is larger. S cannot be increased too greatly by this method. The precipitation must be complete. Most precipitates in quantitative procedures are insoluble enough that a 1000 fold increase in solubility will still allow quantitative, that is, one part per 1000, separation.

4. *Precipitation from a hot solution.* Most precipitates are more soluble at higher temperatures. Since an increase in S decreases the relative supersaturation, the particle size will be larger.

Homogeneous Precipitation

There is an ultimate technique for keeping the degree of supersaturation low. This method is known as *precipitation from a homogeneous solution.* In this procedure the precipitating reagent is generated slowly *in situ* by a chemical reaction which occurs uniformly throughout the solution. The reagent reacts almost as fast as it forms, therefore, local excess concentrations of the reagent do not occur. The number of nuclei which form is relatively small and these nuclei grow as more precipitating reagent is produced. Consequently, relatively few crystals are produced, but they have a large size.

An example of homogeneous precipitation is in the precipitation of hydrated iron(III) oxide. In the classical method, aqueous ammonia is added to an acidic Fe(III) solution until a precipitate forms. The precipitate has the formula $Fe_2O_3 \cdot xH_2O$ with x being a very large and indefinite value. The slow generation of hydroxide ions for the precipitation can be generated by the hydrolysis of urea.

$$(NH_2)_2CO + 3H_2O \rightarrow CO_2 + 2NH_4^+ + 2OH^-$$

At a temperature near 100°C this reaction proceeds slowly. In a typical determination one to two hours are required to complete the precipitation. The precipitate formed by this procedure still has a formula $Fe_2O_3 \cdot xH_2O$ but now x is much smaller. The precipitate formed by this method occupies from one-tenth to one-twentieth the volume of that formed by the classical procedure. As a result, the precipitate is much easier to wash and filter. Also, the amount of impurities coprecipitated is considerably smaller. For example, Gordon* reported that the contamination in the precipitation of 0.1 g of aluminum in a solution containing 1.0 g of manganese amounted to 1.2 mg when precipitated by the classical method and only 0.2 mg when precipitated by homogeneous precipitation.

Methods have been developed for the homogeneous generation of a number of precipitating reagents listed in Table 4-1.

Table 4-1

Reagents Used for Homogeneous Precipitations

Precipitation ion	Hydrolytic reagent
OH^-	Urea, NH_2CONH_2
$C_2O_4^{-2}$	Diethyloxalate, $(C_2H_5)_2C_2O_4$
PO_4^{-3}	Trimethylphosphate, $(CH_3)_3PO_4$
SO_4^{-2}	Sulfamic acid, NH_2SO_3H
S^{-2}	Thioacetamide, CH_3CSNH_2
Oxinate$^-$	8-hydroxyquinoline acetate

Coprecipitation

Precipitates tend to carry down from the solution impurities which under the existing conditions should be soluble. This process is known as coprecipitation. In order to perform a successful gravimetric analysis, the amount of coprecipitation must be kept low. The amount of coprecipitation varies directly with the surface area of the precipitate. Contamination of the precipitate is the most important reason for keeping the degree of supersaturation as low as possible. The lower the supersaturation—the lower the rate of nucleation—the larger the particles—the lower the surface area—the *purer* the precipitate! It is not always possible to completely eliminate coprecipitation, but by careful precipitation and thorough washing, it is possible to minimize its effect. There are four main ways in which coprecipitation can occur.

1. Surface adsorption. Nearly all inorganic precipitates are crystalline in character. They consist of an orderly arrangement of ions

---

*L. Gordon, *Anal. Chem.*, **24**, 459 (1952).

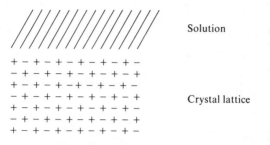

Figure 4-1

Cross sectional view of a simple crystal.

known as a lattice structure. A cross sectional diagram of a crystal such as sodium chloride is given in Figure 4-1; + represents the positive ions, − the negative ions. A positive ion located in the interior of the crystal is surrounded by six equally spaced negative ions, the four located north, south, east, and west and the one directly above and directly below the cross section. However, a positive ion located at the surface only has five negative ions surrounding it. The positive ion at the surface, therefore, has an attractive force towards negative ions in the solution. Likewise, negative ions at the surface of the precipitate will attract positive ions in the solution. The ions which are most easily adsorbed are the ions of the solution which are common to the precipitate. The attractive forces between an ion of the precipitate and its common ion are stronger than the forces between the precipitate ion and another foreign ion. For example, in the precipitation of a chloride sample with silver nitrate, at any time prior to the time that an equivalent amount of silver nitrate has been added, an excess of chloride ions occurs in the solution. Therefore, the crystal of silver chloride will preferentially adsorb chloride ions. This layer of adsorbed ions is known as the *primary adsorbed layer*. The primary adsorbed layer being negatively charged will in turn attract positive ions, e.g., sodium ions, from the solution. This layer of ions is known as the counter ion layer. The primary adsorbed ions are held much more tightly than the counter ion layer. Figure 4-2 shows a cross sectional diagram of a precipitate of silver chloride resulting from the incomplete precipitation of sodium chloride with silver nitrate.

After an excess of silver nitrate has been added, silver is the primary adsorbed ion and nitrate is the counter ion. In order to

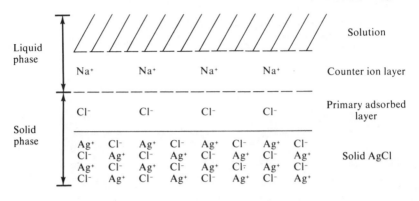

Figure 4-2

Cross sectional diagram of a silver chloride particle
in a solution containing an excess of chloride ions.

prevent high results from occurring, the primary ion layer and
counter ion layer must be removed by washing or replaced by ions
that can be readily volatilized.

2. Occlusion. Occlusion is the trapping of foreign ions or solvent
    within the crystal structure of the precipitate. This type of copre-
    cipitation occurs as the crystal grows. It may occur by two dif-
    ferent mechanisms.

    Mechanical entrapment occlusion occurs when irregular crystal
    growth results in the surrounding of a small amount of mother
    liquor containing dissolved impurities. This always causes high
    results because even if drying the precipitate removes the solvent,
    the dissolved impurities will remain in the crystal. Figure 4-3 is a
    two dimensional illustration of mechanical entrapment occlusion.

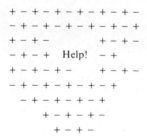

Figure 4-3

Mechanical entrapment occlusion.

The second type of occlusion is surface adsorption occlusion. As mentioned above, prior to stoichiometric completion, a precipitate contains a primary adsorbed ion common to the precipitate and a counter ion foreign to the precipitate. As additional precipitating reagent is added, the counter ion is displaced by the appropriate ion of the reagent. However, if crystal growth is too rapid, the crystal may grow completely around a counter ion trapping it within the crystal. Surface adsorption occlusion can lead to either high or low results depending on whether the occluded "counter ion" is heavier or lighter than the precipitating ion.

3. Postprecipitation. Most precipitates increase in purity on continued digestion in the mother liquor. Sometimes however, when the precipitate is allowed to stand, a second substance will slowly form a precipitate with the precipitating reagent. The second precipitate which forms would not precipitate under identical conditions if it were not for the presence of the first precipitate, because of the tendency for the second precipitate to form a supersaturated solution. This phenomenon is known as postprecipitation. When postprecipitation occurs, digestion increases the amount of contamination. Therefore, when postprecipitation is expected, filtration must be made quickly after the initial precipitation is completed. Examples of postprecipitation are the forming of magnesium oxalate onto a precipitate of calcium oxalate and the forming of zinc sulfide onto a precipitate of cadmium, copper, or mercury(II) sulfides.

In a strict sense postprecipitation is not a form of coprecipitation because it occurs after the desired precipitate is formed. The net result remains the contamination of a precipitate with a substance normally soluble under the existing conditions.

4. Isomorphous replacement. Two compounds are known as isomorphous if they have the same type of formula and crystallize in similar geometric forms. When their lattice dimensions are about the same, one compound can replace parts of the other in a crystal. The result is the formation of mixed crystals. In the precipitation of magnesium ions as magnesium ammonium phosphate, $MgNH_4PO_4$, potassium ions have nearly the same ionic size as the ammonium ions and can replace some of them to form magnesium potassium phosphate, $MgKPO_4$.

Isomorphous replacement, when it occurs, is very serious and little can be done to avoid it unless the offending ion can be removed prior to precipitation. Fortunately, it rarely occurs in analytical precipitates.

Double Precipitation

Sometimes the best laid plans of mice and men go astray. Gravimetric analysis is no exception. In certain cases serious coprecipitation errors cannot be avoided regardless of the skill of the analyst. In these cases a double precipitation becomes necessary. The precipitate containing its impurities is removed from the solution, the precipitate is washed as usual, the washings and filtrate are discarded, the precipitate is dissolved, and the precipitation is made a second time. The concentration of coprecipitable material is now much lower than initially present and consequently the extent of coprecipitation is less. In most cases the amount of impurity carried down in the second precipitation is small enough to be neglected. If the purity of the precipitate is still not adequate, a third precipitation or more will be necessary.

Selectivity of Ion Adsorption

The adsorption of ions on a precipitate is based upon electrostatic attractions. The tendency to preferentially adsorb certain ions is a combination of four factors.

1. Solubility effect. When other factors are equal, the ion which forms the least soluble compound with one of the lattice ions is preferentially adsorbed. As a consequence, when either lattice ion is present in excess in the mother liquor, that ion is preferentially adsorbed. This factor is known as the Paneth-Fajans-Hahn adsorption rule. It also applies to the adsorption of ions when neither lattice ion is available as an excess.
2. Concentration effect. When other factors are equal, the ion present in the greatest concentration is preferentially adsorbed. Also the amount of ion adsorbed is directly proportional to its concentration.
3. Ionic charge effect. When other factors are equal, the larger the ionic charge, the greater the adsorption.
4. Ionic size effect. When other factors are equal, the ion most nearly the size as the lattice ion is preferentially adsorbed.

A combination of these four factors determines what adsorption will occur in a particular situation. The solubility effect is generally the dominant factor, but under certain conditions, any of the four factors may predominate.

Digestion of the Precipitate

Very small crystals have a much higher surface area-to-mass ratio. Therefore, surface adsorption on small crystals is greater than it is on larger crystals. Table 4-2 shows the comparison of surface area on cubes of different sizes.

Table 4-2

Surface Area of Cubes Having a Total
Volume of 1 cm^3

Side of cube, cm	Total surface area, cm^2
1	6
0.1	60
0.01	600

Small crystals have a higher surface energy and are more soluble than large crystals. When a precipitate is allowed to stand in the presence of its mother liquor, the large crystals grow at the expense of the smaller crystals. This process is known as *digestion*. The small crystals dissolve and are reprecipitated on the surface of the larger crystals. Small imperfections in the surface of the crystals also tend to rearrange, freeing some occluded material. Adsorbed impurities on the smaller crystals go back into solution as the crystals dissolve. Digestion is usually done at elevated temperatures to hasten the dissolving and reprecipitating process, although in some cases it is done at room temperature. The net result is larger, more perfect crystals, with fewer impurities.

Some precipitates are so insoluble that $Q - S$ cannot be held small enough to yield a crystalline precipitate. In this case the precipitate when first formed is colloidal. A colloidal particle has a size between 1 and 100 nm in diameter and is too small to be visible even under a light microscope. This size is much too small to be collected on ordinary filter media. Heating a colloidal suspension or adding an electrolyte to it causes the suspension to be *coagulated* into a solid mass known as a colloidal precipitate. Such a colloidal precipitate can be collected in a filter.

The surface of a colloid is charged by the presence of the adsorbed primary and counter ions. The stability of a colloidal suspension is due to the repulsive forces set up among the double layer of adsorbed ions on the individual particles. The charges on the double layer prevents the close approach necessary for the cohesive forces between the particles to bring about coagulation.

The coagulation process involves a reduction in the forces of repulsion caused by the charged double layer. One way to accomplish this is to reduce the total charge of the double layer by reducing the number of ions adsorbed on the particles. Heating the colloidal suspension does this. Thus many colloidal suspensions can be coagulated by heat alone.

A more effective way of causing coagulation is to add an electrolyte to the solution. This tends to minimize the effective charge of the double layer. With sufficiently high electrolyte concentrations, the effect of the double layer is overcome and the particles come close enough into contact with one another to adhere.

There are two types of colloids. Hydrophobic colloids have little if any attraction for water, are coagulated fairly easily, and result in a curdy precipitate. An example of a hydrophobic colloid is silver chloride. Even though silver chloride precipitates as a colloid, it is sufficiently coagulated after one half hour at 80°C to allow filtration. And the gravimetric determination of chloride is one of the most accurate determinations known!

Hydrophilic colloids have a strong attraction for water, are more difficult to coagulate, and the coagulated precipitates are gelatinous and difficult to filter. Examples of hydrophilic colloids are the hydrous metal oxides. The gelatinous nature of these precipitates makes them adsorb impurities easily and this type of gravimetric determination is not as accurate as most gravimetric methods.

Filtration and Washing of the Precipitate

Coprecipitated impurities caused by surface adsorption can be removed by washing the precipitate. Washing also removes the mother liquor which adheres to the surface. Many precipitates cannot be washed with distilled water because *peptization* will occur. Peptization is the breaking up of coagulated particles, forming colloidal particles which pass through the filter. To prevent peptization from occurring, the precipitate is washed with a dilute electrolyte solution.

The electrolyte used for washing must be volatile at the temperature used for drying or ignition. For example, in the gravimetric chloride determination, dilute nitric acid is used as a wash solution. The nitric acid replaces the adsorbed double layer of ions on the precipitate and is volatilized when the precipitate is dried at 100°C.

$$AgCl:Ag^+ \, anion^-(s) + HNO_3 \longrightarrow AgCl:H^+NO_3^-(s) + Ag^+ + anion^-$$

$$AgCl:H^+NO_3^-(s) \xrightarrow[110°]{\Delta} AgCl(s) + HNO_3(g)$$

The most effective washing of a precipitate occurs in the beaker. Very little exchange of double layer ions occurs once the precipitate has been transferred into the filter.

A precipitate may be filtered through paper, a gooch crucible with an asbestos mat, or a sintered glass filtering crucible. The selection of the filtering medium depends on the type of precipitate and the method of drying or igniting the precipitate.

Drying or Igniting the Precipitate

After a precipitate has been filtered and washed, it must be heated to remove water and to volatilize the adsorbed electrolyte from the wash liquid. If the precipitate is in a form suitable for weighing, the drying can usually be done in a drying oven at 110–120°C for one to two hours. Most precipitates can be dried indefinitely at this temperature. Leaving a precipitate in a drying oven overnight or longer ensures that it has been dried to constant weight. Not all precipitates, particularly those containing organic portions, can be dried for extended periods of time.

Sometimes it is desirable to convert the precipitate into a more suitable form for weighing. This process is known as *ignition* and is done in an electric muffle furnace. Examples of ignition of precipitates to a more weighable form include decomposition of magnesium ammonium phosphate, $MgNH_4PO_4$, and zinc ammonium phosphate, $ZnNH_4PO_4$, to their respective pyrophosphate salts, $Mg_2P_2O_7$ and $Zn_2P_2O_7$. These decompositions occur at 1100°C. Hydrous ferric oxide, $Fe_2O_3 \cdot xH_2O$, is converted into anhydrous ferric oxide at 800–1000°C.

The determination of calcium by precipitation with oxalate forms an interesting study on the heating of precipitates. The following weighing forms have been proposed: $CaC_2O_4 \cdot H_2O$, CaC_2O_4, $CaCO_3$, and CaO. Methods are available for conversion of the calcium oxalate into $CaSO_4$ and CaF_2.

Several workers have proposed weighing the precipitate as $CaC_2O_4 \cdot$

H_2O after drying at 100–110°C. However, the dried precipitate tends to contain foreign water and also to be hygroscopic.

At approximately 200°C calcium oxalate monohydrate loses its water of crystallization. The resulting precipitate has been recommended as a weighing form. However, complete removal of water is difficult and the anhydrous calcium oxalate is very hygroscopic.

If the precipitate is ignited within the range of 475–525°C, calcium oxalate decomposes into calcium carbonate which is a very suitable form for weighing. Temperature control must be closely adhered to because calcium carbonate begins to lose carbon dioxide above 550°C.

Above 900°C (preferably around 1100°C), calcium carbonate is converted stoichiometrically into calcium oxide which can be weighed as such. However, calcium oxide also is hygroscopic and appropriate precautions must be taken.

By treating calcium oxalate, carbonate, or oxide with excess sulfuric acid or hydrofluoric acid, evaporating the acid, and igniting the residue, anhydrous calcium sulfate or anhydrous calcium flouride are obtained which can be weighed as such.

Taking proper care any of the above forms can yield accurate results. It should be clear that the effect of heat on a precipitate must be known.

Weighing the Precipitate

After drying the precipitate to constant weight, the crucible must be placed in a desiccator until it is cooled to room temperature. The crucible and precipitate should be weighed rapidly to prevent moisture from being adsorbed.

Calculation of the Results

Calculation of results are normally based on the percentage of an element or ion in the sample.

$$\% A = \frac{\text{wt of } A}{\text{sample wt}} \times 100$$

The weight of the substance being determined is calculated from the weight of the precipitate by means of a *gravimetric factor* or a *chemical factor.* A chemical factor is the ratio of the formula weights of the substance sought to the substance weighed times a stoichiometry factor S.

$$\text{chemical factor} = \frac{\text{formula wt of substance sought}}{\text{formula wt of substance weighed}} \times S$$

S is the ratio of the number of formula weights of the substance sought to the number of formula weights of the substance weighed that occurs in the balanced chemical equation(s).

For example, in the gravimetric determination of chloride, one formula weight of silver chloride is formed for each formula weight of chloride present.

$$Ag^+ + Cl^- \rightarrow AgCl$$

S is unity and the chemical factor for chloride in silver chloride is given by the following:

$$\text{chemical factor} = \frac{\text{formula wt of Cl}}{\text{formula wt of AgCl}} = \frac{35.45}{143.32} = 0.2474$$

It is troublesome to include the words "formula weight of" in calculations. In this book whenever a chemical symbol appears in a calculation, it means formula weight of that chemical. The above chemical factor simplifies to

$$\text{chemical factor} = \frac{Cl}{AgCl} = 0.2474$$

By multiplying any weighed amount of silver chloride precipitate by 0.2474, the weight of chloride in the precipitate is obtained. The percentage of chloride in a sample is calculated as follows:

$$\% \, Cl^- = \frac{\text{wt of AgCl} \times \dfrac{Cl^-}{AgCl} \times 100}{\text{sample wt}}$$

One can analyze this equation in the following manner. Divide the weight of precipitate by the formula weight for silver chloride. This yields the number of formula weights of silver chloride in the precipitate:

$$\frac{\text{wt of AgCl}}{\text{AgCl}} = \text{number of formula weights of AgCl}$$

Referring to the balanced chemical equation, it is seen that one silver chloride is obtained from one chloride. Therefore, the number of formula weights of silver chloride equals the number of atomic weights of chloride:

$$\frac{\text{wt of AgCl}}{\text{AgCl}} = \text{number of atomic weights of Cl}^-$$

Multiplying the number of atomic weights of chloride by the atomic weight of chlorine yields the weight of chloride:

$$\frac{\text{wt of AgCl}}{\text{AgCl}} \times \text{Cl} = \text{wt of Cl}^-$$

And the percentage is found directly:

$$\frac{\text{wt of Cl}}{\text{sample wt}} \times 100 = \%\text{Cl}^-$$

This latter equation is usually written in the form

$$\frac{\text{wt of AgCl} \times \dfrac{\text{Cl}}{\text{AgCl}} \times 100}{\text{sample wt}} = \%\text{Cl}^-$$

In general, the calculation of percentage of A in a sample is given by the following:

$$\frac{\text{wt of precipitate} \times \text{chemical factor} \times 100}{\text{sample wt}} = \%A$$

The chemical factor expresses the ratio of weights of two substances that are equivalent to each other. Therefore, *almost without exception* the same number of atoms of the constituent concerned must appear in the numerator and denominator. Table 4-3 lists some common examples.

Often the substance sought does not occur in the substance weighed. As long as the stoichiometric relation between the two substances is known the chemical factor can be calculated.

Table 4-3

Chemical Factors

Substance weighed	Substance sought	Factor
Fe_2O_3	Fe	$\dfrac{2Fe}{Fe_2O_3}$
Fe_2O_3	Fe_3O_4	$\dfrac{2Fe_3O_4}{3Fe_2O_3}$
$Mg_2P_2O_7$	PO_4^{-3}	$\dfrac{2PO_4}{Mg_2P_2O_7}$
$Mg_2P_2O_7$	P_2O_5	$\dfrac{P_2O_5}{Mg_2P_2O_7}$
$Mg_2P_2O_7$	MgO	$\dfrac{2MgO}{Mg_2P_2O_7}$

To determine the amount of sodium in a solution of pure sodium chloride, the chloride can be precipitated as silver chloride and the correct factor is used to calculate the weight of sodium in the sample.

$$NaCl \longrightarrow Na^+ + Cl^-$$

$$Cl^- + AgNO_3 \longrightarrow AgCl + NO_3^-$$

$$\text{chemical factor} = \frac{Na}{AgCl}$$

No matter how many intermediate compounds are involved in a series of reactions, only the stoichiometry between first and last need be considered in setting up the chemical factor.

EXAMPLE 4-4: A sample containing sodium bromide is treated with silver nitrate solution. The resulting silver bromide precipitate is converted into silver chloride with chlorine gas and the resulting silver chloride is weighed.

$$Na^+ + Br^- + Ag^+ \rightarrow AgBr + Na^+$$

$$2AgBr + Cl_2 \rightarrow 2AgCl + Br_2$$

One formula weight of sodium bromide is converted into one formula weight of silver chloride by the reactions, therefore,

$$\text{chemical factor} = \frac{NaBr}{AgCl}$$

Chemical factors are not unique to gravimetric analysis. They are also used to express the relation between weights of substances in any type of reaction which proceeds to completion. Consider the following reactions:

$$2KMnO_4 + 10KI + 16HCl \longrightarrow 12KCl + 2MnCl_2 + 5I_2 + 8H_2O$$

$$2Na_2S_2O_3 + I_2 \longrightarrow 2NaI + Na_2S_4O_6$$

The factor for $Na_2S_2O_3$ in $KMnO_4$ is given by

$$\frac{10Na_2S_2O_3}{2KMnO_4} = 5.003$$

For each gram of potassium permanganate involved in the first reaction, 5.003 g of sodium thiosulfate will be required for the second reaction.

An example of a chemical factor in which the number of atoms of the concerned constituent is not identical in the numerator and denominator is shown by the following reaction:

$$MnO_2 + 4HCl \longrightarrow MnCl_2 + Cl_2 + 2H_2O$$

The chemical factor for chlorine gas in hydrogen chloride is given as follows:

$$\text{chemical factor} = \frac{Cl_2}{4HCl}$$

The latter example is indeed an exception. But it is given to illustrate the point that whenever in doubt reference must be made to the chemical equations involved.

Indirect Analysis

When determination of two constituents of a sample are to be made, a separate determination is usually performed on each constituent.

However, it is possible to determine the amounts of two substances from the change in weight that a known amount of the mixture undergoes when converted into other substances which differ in molecular weight both between themselves and from the original substances. This is known as indirect analysis. For example, the chloride ion from a known weight of a mixture of pure sodium chloride and pure potassium chloride can be precipitated as silver chloride. From the weight of precipitate and the chemical factors the composition of the mixture can be determined. The following example illustrates the method:

EXAMPLE 4-5: A sample is known to contain only NaCl and KCl with no impurities. From this sample, 0.6000 g was dissolved and the chloride determined by a gravimetric procedure. After drying, the AgCl precipitate weighed 1.2460 g. Calculate the weight of NaCl and KCl in the sample.

The weights of NaCl and KCl multiplied by their respective chemical factors gives the weight of silver chloride obtained from them.

Let X = weight of NaCl.

$$0.6000 \text{ g} - X = \text{weight of KCl}$$

$$(X \text{ g}) \left(\frac{AgCl}{NaCl}\right) + (0.6000 \text{ g} - X) \left(\frac{AgCl}{KCl}\right) = 1.2466 \text{ g}$$

$$2.452X + (0.6000 - X)(1.922) = 1.2466$$

Combining terms and rearranging gives

$$(2.452 - 1.922)X = 1.2466 - 1.1532$$

$$X = \frac{0.093}{0.530} = 0.176 \text{ g}$$

The weight of NaCl is 0.176 g and the weight of KCl is 0.424 g.

Indirect methods must be used with caution because they usually do not yield highly accurate results. A small error in the determination is compounded several fold in the result. In the above example, an error of 0.1 mg in obtaining the weight of mixed chlorides causes an error of 0.4 mg in the weight of sodium chloride found. Thus, much of the relative accuracy obtained in the measurements is lost in the mathematical operations. Consequently, indirect analysis is generally advisable only when quantitative separation and direct analysis is not feasible.

Questions

1. List the three criteria which a precipitation process must fulfill to make it suitable for a gravimetric determination.

2. An aqueous solution contains sodium sulfate. The sulfate ion is precipitated by the slow addition of a solution of barium chloride. (a) Before sufficient barium chloride is added for precipitation to be complete what is the primary adsorbed ion and the counter ion? (b) After an excess of barium chloride is added what is the primary adsorbed ion and the counter ion?

3. Define the term quartering.

4. Explain the proper use of a desiccator.

5. Write the von Weimarn expression and define all terms.

6. Discuss what experimental factors can be adjusted to increase the particle size of a precipitate.

7. Explain why homogeneous precipitation results in a larger particle size and a more pure precipitate.

8. Occlusion can occur by two methods. Explain how and why each type of occlusion occurs.

9. Discuss the four factors which affect the adsorption of ions on a precipitate.

10. What is digestion? Why does digestion usually improve the quality of a precipitate? When does digestion not result in a more pure precipitate?

11. What is peptization? How may it be avoided?

12. If some sodium chloride is occluded during the precipitation of silver chloride would the results for the percentage of chloride in the sample be high or low?

13. In the filtration of an analytical precipitate, why is it not advisable to transfer the bulk of the precipitate into the filter crucible as soon as possible?

14. Comment on the general statement: In a gravimetric or chemical factor the number of atoms of the principle element must be the same in both the numerator and the denominator.

15. In the gravimetric analysis of chloride by precipitation of silver chloride explain the use of each of the following steps:
    (a) the heating of the solid sample prior to weighing
    (b) the heating of the solution prior to the addition of silver nitrate
    (c) the addition of nitric acid to the solution prior to the addition of silver nitrate
    (d) the use of an excess volume of silver nitrate solution
    (e) the presence of a small amount of nitric acid in the wash solution
    (f) the testing of the filtrate with dilute hydrochloric acid

Problems

1. A sample of soda ash contained 4.24% moisture and 88.9% sodium carbonate on the as received basis. Calculate the percent sodium carbonate in the sample after it is completely dried.

2. A sample of coal lost 3.98% of its weight when dried to constant weight. The dried sample was found to contain 10.4% ash. Calculate the percent ash in the undried coal sample.

3. A paint sample was found to contain 35.2% volatile thinner and 14.4% pigment. The lid was accidentally left off the container for a period of time. The paint was reblended and now was found to contain 27.6% thinner. Calculate the percent pigment now in the paint.

4. An iron ore sample contained 42.85% Fe_2O_3 as received and 46.22% Fe_2O_3 after oven drying. Calculate the percent moisture in the sample.

5. Company A purchased a quantity of material from company B. The invoice from A showed the material to contain 82.5% active ingredient and 2.6% moisture. An analyst working for company B found only 80.4% active ingredient and 4.92% moisture. Calculate the actual absolute difference in percent active ingredient between the analyst's results and the invoice.

6. Calculate the gravimetric factor for the following:
   (a) Pb in Pb_3O_4 (b) Zn in $Zn_2P_2O_7$
   (c) NH_3 in $(NH_4)_2\,PtCl_6$ (d) S in $BaSO_4$
   (e) SO_3 in $BaSO_4$ (f) Ag in AgCl

7. Calculate the chemical factor used for converting the following:
   (a) AgCl to $KClO_3$
   (b) $BaSO_4$ to $Fe(NH_4)_2\,(SO_4)_2 \cdot 6H_2O$

8. Calculate the chemical factor for Cl_2 in HCl according to the following equation.

$$MnO_2 + 4HCl \rightarrow MnCl_2 + Cl_2 + 2H_2O$$

9. Consider the following sequence of balanced equations:

$$KIO_3 + 5KI + 6HCl \rightarrow 3I_2 + 6KCl + 3H_2O$$
$$I_2 + 2Na_2S_2O_3 \rightarrow Na_2S_4O_6 + 2NaI$$
$$NaI + H_3PO_4 \rightarrow NaH_2PO_4 + HI$$

Calculate the weight of HI formed by these reactions from 2.25 g of KIO_3. Assume each reaction takes place to 100%.

10. Calculate the volume of $0.20M$ silver nitrate solution required for a 10% excess for precipitation of the chloride from a 0.5000 g sample of pure sodium chloride.

11. Calculate the weight of silver chloride which can be formed from a 1.025 g sample of $BaCl_2 \cdot 2H_2O$.

12. Calculate the percentage of sulfur in a sample if a 0.5024 g sample yields a precipitate of barium sulfate weighing 0.4264 g.

13. Calculate the percent purity of a sample of ferrous ammonium sulfate $Fe(NH_4)_2(SO_4)_2 \cdot 6H_2O$ if a 1.000 g sample yields 0.1982 g of Fe_2O_3.

14. Calculate the percentage of chloride in a 0.4000 g sample which yields a precipitate of silver chloride weighing 0.9290 g.

15. The calcium from a 0.5000 g sample of limestone was precipitated as calcium oxalate and ignited to calcium carbonate which weighed 0.2760 g. Calculate the percentage of calcium in the sample.

16. Calculate the weight of sample which should be taken for analysis so that the weight of silver chloride when multiplied by 100 will give the percentage of chloride in the sample.

17. A sample is known to contain only sodium chloride and potassium chloride with no other impurities. A 0.6000 g sample is dissolved, and the total chloride is determined by precipitation. After drying, the silver chloride precipitate weighed 1.246 g. Calculate the percentage of sodium chloride in the sample.

18. A sample containing only calcium carbonate and barium carbonate with no other impurities was found to contain the same percentage of carbon dioxide as there is in pure strontium carbonate. Calculate the composition of the sample.

5

Calculations Involving Saturated Solutions of Slightly Soluble Salts

Classical gravimetric analysis involves the separation of a precipitate from a solution which is saturated with the precipitate. It is important to know what factors affect the amount of substance remaining in a saturated solution.

Solubility

The solubility of a substance expresses the concentration of the solute in a given solvent at a given temperature when equilibrium has been reached between the undissolved substance and the solution.

The equilibrium established between a solid and its solution is dynamic as evidenced by the fact that a precipitate increases in purity and in particle size when left in its mother liquor. The time required to reach equilibrium depends upon the temperature and viscosity of the solution, the amount of stirring, and other factors. But in all cases, the rate of change in concentration approaches zero as equilibrium is approached.

89

Solubility Product Constant

In a saturated solution of a slightly soluble salt, the product of the activities of the ions (each activity raised to a power corresponding to the number of such ions occurring in the balanced chemical equation) is a constant known as the *solubility product constant.* This constant holds for a given solvent and at a given temperature.

For a precipitate of silver chloride, the equilibrium constant for the reaction

$$AgCl_{(s)} \rightleftharpoons Ag^+ + Cl^-$$

is

$$K = \frac{a\,Ag^+\,a\,Cl^-}{a\,AgCl_{(s)}}$$

The activity of solid AgCl is constant and by convention is taken as unity. In dilute solutions, the activities of the ions are approached by the molarities of the ions. Therefore, the expression becomes

$$K_{sp} = [Ag^+][Cl^-]$$

Table 5-1 lists the solubility product expressions for a few precipitates. Their numerical value along with that of many more precipitates is given in Appendix B. Solubility product constants are calculated from solubility data.

Table 5-1

Solubility Product Expressions for Selected Precipitates

Precipitate	Expression
AgCl	$K_{sp} = [Ag^+][Cl^-]$
Ag_2CrO_4	$[Ag^+]^2[CrO_4^{--}]$
$Ca_3(PO_4)_2$	$[Ca^{++}]^3[PO_4^{---}]^2$
$MgNH_4PO_4$	$[Mg^{++}][NH_4^+][PO_4^{---}]$

EXAMPLE 5-1: Calculate the solubility product constant for Ag_3PO_4 if 500 ml of saturated solution contains 0.00325 g of dissolved salt. Neglect any hydrolysis effects.

$$Ag_3PO_4 \rightleftharpoons 3Ag^+ + PO_4^{---}$$

$$0.00325 \text{ g}/500 \text{ ml} = 0.00650 \text{ g}/1$$

$$\frac{0.00650 \text{ g}/1}{418.7 \text{ g/m}} = 1.55 \times 10^{-5}M$$

Each mole of silver phosphate which dissolves produces one mole of phosphate ions and *three* moles of silver ions. Therefore,

$$[Ag^+] = 4.65 \times 10^{-5}$$

$$[PO_4^{---}] = 1.55 \times 10^{-5}$$

$$K_{sp} = [Ag^+]^3[PO_4^{---}]$$

$$= (4.65 \times 10^{-5})^3(1.55 \times 10^{-5}) = 1.6 \times 10^{-18}$$

It should be noted that in any equilibrium constant calculation, actual equilibrium concentrations are inserted into the expression.* A common error in the above type problem, when given the silver ion concentration, is to triple its value. The silver ion concentration is never tripled. What actually happens is that when neither ion is in excess the solubility must be tripled to obtain the silver ion concentration. Example 5-2 serves as an illustration.

EXAMPLE 5-2: The silver ion concentration in a saturated solution of silver chromate at 25°C is $1.7 \times 10^{-4}M$. Calculate the solubility product of Ag_2CrO_4 at this temperature.

$$Ag_2CrO_4 \rightleftharpoons 2Ag^+ + CrO_4^{--}$$

$$K_{sp} = [Ag^+]^2[CrO_4^{--}]$$

It is seen by the equation that the chromate ion concentration is one-half as great as the silver ion concentration.

$$[Ag^+] = 1.7 \times 10^{-4}$$

$$[CrO_4^{--}] = 8.5 \times 10^{-5}$$

$$K_{sp} = (1.7 \times 10^{-4})^2(8.5 \times 10^{-5})$$

$$= 2.4 \times 10^{-12}$$

---

*Again for theoretical calculations activities should be used. In this text activities will not be considered in these calculations.

Solubility product constants can also be used to calculate the solubility of a precipitate.

EXAMPLE 5-3: Calculate the solubility of silver chloride at 25°C in moles per liter. The solubility product constant is 1.8×10^{-10}.

$$AgCl_{(s)} \rightleftharpoons Ag^+ + Cl^-$$

$$K_{sp} = [Ag^+][Cl^-] = 1.8 \times 10^{-10}$$

Each mole of silver chloride which dissolves yields one mole of silver ions and one mole of chloride ions. Since there is no other source of silver or chloride ions, $[Ag^+]$ equals $[Cl^-]$ and is equal to the molar solubility of the salt.

Let X = molar solubility = $[Ag^+]$ = $[Cl^-]$.

$$X^2 = 1.8 \times 10^{-10}$$

$$X = 1.3 \times 10^{-5}$$

$$\text{solubility} = 1.3 \times 10^{-5} \text{ m/l}$$

The solubility can be expressed in any weight unit per any given volume unit. The solubility can also be used to express the concentration of an ion in the saturated solution.

EXAMPLE 5-4: Calculate the weight in milligrams of chloride ion remaining in 200 ml of a saturated solution of silver chloride. From Example 5-3 we see that a saturated solution of AgCl contains 1.3×10^{-5} m/l dissolved silver chloride. The solution also contains 1.3×10^{-5} m/l of chloride ion.

$$\text{solubility} = 1.3 \times 10^{-5} \text{ m/l} \times 35.45 \text{ g/m}$$

$$\times 1000 \text{ mg/g} \times \frac{200 \text{ ml}}{1000 \text{ ml/l}}$$

$$= 0.092 \text{ mg/200 ml}$$

Common Ion Effect

Examples 5-1 to 5-4 have all been for solutions of a saturated salt in pure water. In a gravimetric procedure, at equilibrium, an excess of the precipitating ion always exists. According to the Principle of Le Chatelier when a system is in equilibrium and a stress is brought to bear upon the equilibrium, the point of equilibrium will shift in such a way as to relieve the stress. Applying this principle to the silver chloride precipitate shows that an increase in either the silver ion or chloride ion concentration will cause the concentration of the other ion to decrease in such a manner as to satisfy the solubility product constant. The solubility of a salt, in the presence of an excess of one of its ions, is determined by the concentration of the ion or ions not in excess.

EXAMPLE 5-5: Calculate the solubility of silver chloride at 25°C in a solution which at equilibrium has a silver ion concentration equal to $0.010M$.

$$AgCl \rightleftharpoons Ag^+ + Cl^-$$

$$K_{sp} = [Ag^+][Cl^-] = 1.8 \times 10^{-10}$$

The solubility of silver chloride in this solution is equal to the chloride ion concentration because each mole of silver chloride which dissolves furnishes one mole of chloride ions.

$$[Cl^-] = \frac{1.8 \times 10^{-10}}{[Ag^+]}$$

$$= \frac{1.8 \times 10^{-10}}{1.0 \times 10^{-2}} = 1.8 \times 10^{-8}\,m/l$$

Comparison of the answer for Example 5-5 to the answer for Example 5-3 shows that the solubility of AgCl in pure water is almost 1000 times greater than it is when an excess $[Ag^+] = 0.010M$ is provided. Normally an excess of precipitating agent of approximately $0.010M$ is used in a gravimetric analysis. This results in a more quantitative precipitation. A large excess of precipitating agent should be avoided as it may increase the solubility due to the formation of a soluble complex. A large excess will also increase the chance of co-precipitation by surface adsorption.

Note the wording in Example 5-5. It specifies that the silver ion concentration at equilibrium is equal to $0.010M$. Technically this is not the same as saying the precipitate is in, for example, $0.010M$ silver nitrate solution. Worded in the latter manner requires a slightly different approach to the solution as follows:

Let X = solubility.

$$[Cl^-] = X$$
$$[Ag^+] = 0.010 + X$$

and
$$(X)(0.010 + X) = 1.8 \times 10^{-10}$$

This is a quadratic equation. However, if the assumption is made that X is much smaller than 0.010, the solution to this problem is identical to the above solution. A quick check shows that X is indeed small compared to $0.010M$ and, therefore, the assumption is justified. This proves to be the usual case and either wording is suitable for the problem.

Diverse Ion Effect

Experimentally, it is found that a precipitate becomes more soluble when a salt which has no ions common to the precipitate is added to the solution. The solubility product is strictly a constant only in extremely dilute solutions. Deviations from it increase as the concentrations of ions in the solution increases. The presence of a salt not common to the ions of the precipitate increases the ionic strength of the solution and consequently lowers the activity coefficient and the activity of the ions. A decrease in activity results in an increase in solubility.

Suitability of a Precipitate for Gravimetric Analysis

In Chapter 4 one of the criteria for gravimetric analysis was quantitative separation. As seen in Example 5-5 an increase in common ion concentration leads to a more complete precipitation. The presumption should not be made, however, that all precipitates are insoluble enough for quantitative precipitation. The following example serves to illustrate the considerations required to determine the feasibility of a precipitate for gravimetric analysis.

EXAMPLE 5-6: The K_{sp} of $AgBrO_3$ is 6.6×10^{-5}. Considering an excess $[BrO_3^-]$ of $0.10M$, is the precipitate suitable for the gravimetric determination of Ag^+? Assume 0.6000 g of silver in the sample.

$$AgBrO_3 \rightleftharpoons Ag^+ + BrO_3^-$$

$$K_{sp} = [Ag^+][BrO_3^-] = 6.6 \times 10^{-5}$$

The concentration of silver ion remaining in the solution is

$$[Ag^+] = \frac{K_{sp}}{[BrO_3^-]}$$

$$= \frac{6.6 \times 10^{-5}}{0.10} = 6.6 \times 10^{-4} M$$

The weight of silver left in 100 ml of solution is

$$\text{grams} = (6.6 \times 10^{-4}\,\text{m/l})(107.88\,\text{g/m})\,\frac{100\,\text{ml}}{1000\,\text{ml/l}}$$

$$= 0.0071\,\text{g}$$

This does not constitute quantitative precipitation. Of the 0.6000 g of silver originally present, the weight remaining represents

$$\frac{0.0071}{0.6000} \times 100 = 1.18\%$$

To be a quantitative precipitation at least 99.9% must be removed. Therefore, bromate ion is not a suitable precipitating agent for silver.

Fractional Precipitation

When a solution contains two ions capable of being precipitated with a certain reagent and the reagent is added slowly to the solution, the substance with the lower solubility will normally precipitate first. The amount of this substance remaining in solution when the second substance just starts to precipitate can be calculated from the respective solubility product constants. If more than 99.9% of the first substance has been removed, a procedure for the determination of the two substances can possibly be developed.

EXAMPLE 5-7: A solution contains $0.10M$ Cl^- and $0.10M$ CrO_4^{--}. Dilute $AgNO_3$ is added slowly while stirring. Calculate which precipitate, $AgCl$ or Ag_2CrO_4, will form first and calculate the concentration of the first ion remaining in solution when the second ion starts to precipitate. $K_{sp}(AgCl) = 1.8 \times 10^{-10}$; $K_{sp}(Ag_2CrO_4) = 2.4 \times 10^{-12}$.

First calculate the $[Ag^+]$ required to precipitate each ion if only it were in the solution.

$$AgCl \rightleftharpoons Ag^+ + Cl^-$$

$$[Ag^+] = \frac{1.8 \times 10^{-10}}{0.10} = 1.8 \times 10^{-9}M$$

$$Ag_2CrO_4 \rightleftharpoons 2Ag^+ + CrO_4^{--}$$

$$[Ag^+]^2 = \frac{2.4 \times 10^{-12}}{0.10}$$

$$[Ag^+] = 4.9 \times 10^{-6}M$$

Obviously, if silver nitrate is added slowly, a silver ion concentration of $1.8 \times 10^{-9}M$ will be attained before $4.9 \times 10^{-6}M$ is attained. Thus, silver chloride will form first. The problem now becomes, "What is the chloride ion concentration left in solution when silver chromate starts to precipitate?" To answer this, one calculates the equilibrium chloride ion concentration when the silver ion concentration has risen to $4.9 \times 10^{-6}M$ (the concentration required just to start the precipitation of Ag_2CrO_4).

$$[Ag^+][Cl^-] = 1.8 \times 10^{-10}$$

$$[Cl^-] = \frac{1.8 \times 10^{-10}}{4.9 \times 10^{-6}} = 3.7 \times 10^{-5}M$$

At the instant Ag_2CrO_4 starts to precipitate, the chloride ion has been reduced to $3.7 \times 10^{-5}M$.

To determine if the chloride ion has been quantitatively removed, calculate the percentage remaining unprecipitated.

$$\% \text{ unprecipitated} = \frac{3.7 \times 10^{-5}}{0.10} \times 100 = 0.037\%$$

The precipitation is quantitative and in fact forms the basis for the Mohr titration of chloride described in Chapter 10.

Effect of Complex Ion Formation on Solubility

The solubility of precipitates is greatly enhanced when in the presence of a complexing agent which can react with the cation of the precipitate. Evidence of this behavior is the complete dissolution of AgCl in excess ammonia solution.

$$AgCl_{(s)} + 2NH_3 \rightleftharpoons Ag(NH_3)_2^+ + Cl^-$$

Many other examples are utilized in analytical chemistry to prevent the precipitation of interfering substances. Reference to more advanced textbooks will disclose the technique for calculating solubilities when complex ion formation is involved.

Effect of pH on Solubility

Certain precipitates act as proton acceptors and, by an acid-base reaction, are converted into soluble species. Examples are the insoluble metal hydroxides and the insoluble salts of weak acids such as carbonates, sulfides, phosphates, oxalates, and so on. By controlling the pH, separations of metal ions can be effected. Again the details of these calculations will be left to a more rigorous course.

Questions

1. Explain why silver chloride is more soluble in a solution containing dilute sodium nitrate than it is in water but is less soluble in a dilute solution of sodium chloride than in water.

2. Explain why a large excess of a common ion is not desirable in a gravimetric analysis.

3. Explain why calcium oxalate is insoluble in a basic solution but is soluble in an acidic solution.

4. Explain why silver chloride, AgCl, $(K_{sp} = 1.8 \times 10^{-10})$ is more soluble than silver cyanide, AgCN, $(K_{sp} = 2 \times 10^{-12})$ but is less soluble than silver chromate, Ag_2CrO_4, $(K_{sp} = 2.4 \times 10^{-12})$.

5. What calculations need to be made to determine whether a precipitate is theoretically satisfactory for a gravimetric analysis?

6. What calculations need to be made to determine whether two substances can be separated by fractional precipitation?

Problems

1. Set up expressions for the solubility products of the following substances:
   (a) $Al(OH)_3$ (b) $Ba_3(AsO_4)_2$
   (c) CaF_2 (d) PbC_2O_4
   (e) Hg_2Cl_2 (f) Tl_2S

2. Calculate the solubility product constants for the following species from the given solubilities:
   (a) AgI, 0.00235 mg/l
   (b) $Ag_2C_2O_4$, 3.28 mg/100 ml

3. A salt has the formula A_2B_3 and a molecular weight of 250 g/mole. The solubility of the salt is 0.0252 mg/100 ml. Calculate the solubility product constant for the salt.

4. The solubility of lead sulfate is 4.25 mg per 100 ml at 25°C. Calculate the following:
   (a) K_{sp} at 25°C
   (b) the molar solubility in 0.010M Na_2SO_4

5. The K_{sp} for $Ba(IO_3)_2$ is 1.5 $\times 10^{-9}$. Calculate, for a saturated solution, the following:
   (a) the molar solubility
   (b) the grams of dissolved $Ba(IO_3)_2$ in 250 ml of solution
   (c) the concentration of IO_3^- in solution

6. The K_{sp} for Ag_2CrO_4 is 2.4 $\times 10^{-12}$. Calculate the following:
   (a) the molar solubility in 0.010M $AgNO_3$
   (b) the molar solubility in 0.010M Na_2CrO_4

7. A salt has the formula A_3B, a molecular weight of 250 g/mole, and a K_{sp} of 4.3 $\times 10^{-16}$. Calculate the molar solubility of the salt.

8. The solubility product constant for $MgNH_4PO_4$ is 2.5 $\times 10^{-13}$. Calculate the molar solubility of the salt in water. Ignore any effects due to hydrolysis.

9. The solubility of Ag_3PO_4 is 6.5 $\times 10^{-3}$ grams per liter at 20°C. Calculate the following:
   (a) K_{sp} for Ag_3PO_4 at 20°C
   (b) the molar solubility in 0.010M $AgNO_3$ solution

10. The solubility product constant for magnesium fluoride, MgF_2, is 7 $\times 10^{-9}$. Calculate the following:
    (a) the molar solubility of MgF_2 in pure water
    (b) the molar solubility of MgF_2 in 0.01M $Mg(NO_3)_2$ solution
    (c) the molar solubility of MgF_2 in 0.01M NaF

11. Calculate the weight in milligrams of $BaSO_4$ which can remain dissolved in 250 ml of a solution which at equilibrium contains 0.05M Ba^{++}.

12. The solubility product constant for A_2B is 3.2×10^{-13}. Show by appropriate calculations whether a solution prepared by mixing equal volumes of $2.5 \times 10^{-4}M$ ACl with $1.0 \times 10^{-5}M$ Na_2B will or will not cause a precipitate to form.

13. Show by appropriate calculations whether SO_4^{--} can be used for the quantitative determination of silver. K_{sp} for Ag_2SO_4 is 1.6×10^{-5}. Assume 100 ml of sample containing $0.10M$ Ag^+. Also assume sufficient SO_4^{--} has been added to give an equilibrium concentration of $0.042M$.

14. Silver nitrate is added to a $0.00010M$ solution of hydrochloric acid until the excess silver ion concentration is $0.010M$. Calculate whether the precipitation of silver chloride under these conditions is quantitative.

15. A solution is $0.10M$ in strontium ion and $0.05M$ in calcium ion. If solid Na_2CO_3 is slowly added to the solution, calculate which carbonate salt will precipitate first and what percentage of it remains unprecipitated when the second carbonate salt just begins to precipitate. K_{sp} for $SrCO_3$ is 2×10^{-9} and for $CaCO_3$ is 3×10^{-14}.

16. A solution is $0.05M$ in lead ion and $0.02M$ in chromium ion. Calculate the hydroxide ion concentration required just to start the precipitation of $Pb(OH)_2$ and $Cr(OH)_3$ respectively. K_{sp} for $Pb(OH)_2$ is 1.2×10^{-15} and for $Cr(OH)_3$ is 6.0×10^{-31}.

17. The K_{sp} for AgCl is 1.8×10^{-10} and K_{sp} for AgBr is 5.2×10^{-13}. A solution contains $0.26M$ NaCl and $0.02M$ KBr. A dilute $AgNO_3$ solution is added while stirring. Show by appropriate calculations which precipitate will form first. Calculate the percentage of the first ion remaining in the solution when the second ion just starts to precipitate. Assume no change in volume during precipitation. Is the separation quantitative?

18. A solution is $0.10M$ in chloride and $0.10M$ in iodide ion. K_{sp} for AgCl is 1.8×10^{-10}, K_{sp} for AgI is 8.3×10^{-17}. Dilute $AgNO_3$ solution is added while stirring. Show as in problem 17 if this is a quantitative separation.

19. Calculate the maximum value of K_{sp} a salt of the type AB can have in order to be used for the quantitative precipitation of A. Assume that the excess of B at equilibrium is $0.035M$ and that the original concentration of A is $0.20M$ (corrected for dilution effect upon precipitation).

20. Calculate the maximum value of K_{sp} a salt of the type A_2B can have in order to be used for the quantitative precipitation of A. Assume that the excess of B at equilibrium is $0.035M$ and that the original concentration of A is $0.20M$ (corrected for dilution effect upon precipitation).

6

Volumetric Analysis

In a volumetric analysis, the volume of a solution of known concentration required to react with the desired constituent is measured. The process of determining the volume is known as a titration and the method is often referred to as titrimetric analysis. Volumetric analysis is much more rapid than gravimetric analysis. Although not quite as accurate as gravimetric methods, accuracy to ± 1–3 parts per thousand is easily attainable in many standard procedures.

Fundamental Principle of a Titration

The fundamental principle upon which the calculations of volumetric analysis are based is as follows: At the equivalence point the number of equivalent (or milliequivalent) weights of substance A being titrated is equal to the number of equivalent (or milliequivalent) weights of substance B in the titrating solution.

101

Requirements of a Titration

Of the thousands of known chemical reactions, only a small percentage can be used for the basis for titrations. In order for a reaction to be utilized in a titration the following requirements must be met:

1. The reaction must proceed in a single, stoichiometric manner. There must not be side reactions, because if there were, then calculations based upon equivalent weights would be impossible.
2. The reaction must be complete and rapid. The equilibrium must lie far to the right so that a large, abrupt change in the concentration in the substance titrated will occur when an equivalent amount of titrant is added. The reaction must be rapid, so that the titration may be completed in a reasonable period of time. If the reaction is complete and stoichiometric, but equilibrium is attained slowly, a back-titration method may be utilized.
3. A means of detecting the equivalence point must be available. If the first two requirements are met, but no means are available to determine the large, abrupt change in concentration occurring at the equivalence point, the reaction cannot serve as a titration method. For many titrations, chemical substances called indicators are used which change color when excess titrant is present. Changes in electrochemical or physical properties can also be used to detect when excess titrant is added. The point at which the indicator changes color is known as the *end point*. The point at which the *theoretical amount* of titrant has been added is known as the *equivalence point*. Ideally the two points should coincide. Slight differences in them can be corrected by the standardization procedure.

Correction of End Point Error

The end point error in a titration is the difference between the equivalence point and the end point. Many titrations have no end point error or a negligible end point error. The nature of other titrations is such that a significant titration error would occur if no correction were applied.

The easiest method for eliminating an end point error is to standardize the titrant in exactly the same manner that it will be used in the analysis. For example, the end point in the titration of an unknown chloride sample with standard silver nitrate always occurs slightly past the equivalence point. To eliminate this error, the silver nitrate should be standardized by titration against a known weight of primary standard sodium chloride duplicating the conditions which will be used in the unknown determination. If the sample weights are adjusted such that the primary standard and the unknown each require approximately the same volume of silver nitrate, the end point error will be canceled because the normality of the silver nitrate will be slightly low in the standardization but the volume used for the unknown will be slightly high.

Types of Titration Reactions

Chemical reactions which serve as the basis for volumetric determinations can be grouped into four types with examples as follows:

1. Acid-base, or neutralization

$$H_3O^+ + OH^- \rightleftharpoons 2H_2O$$

or simply

$$H^+ + OH^- \rightleftharpoons H_2O$$

2. Oxidation-reduction (redox)

$$6Fe^{++} + Cr_2O_7^{--} + 14H^+ \rightleftharpoons 6Fe^{+++} + 2Cr^{+++} + 7H_2O$$

3. Precipitation

$$Ag^+ + Cl^- \rightleftharpoons AgCl$$

4. Complex formation

$$Ag^+ + 2CN^- \rightleftharpoons Ag(CN)_2^-$$

Unit of Volume

The fundamental unit of volume is the *liter*, defined as the volume occupied by 1 kilogram (vacuum weight) of water at 3.98°C and 1 atmosphere of pressure. The milliliter is defined as below:

$$1 \text{ milliliter} = \frac{1}{1000} \text{ liter}$$

The milliliter is more convenient to use in volumetric analysis than the liter.

Another unit of volume is the cubic centimeter. The milliliter and cubic centimeter can be used interchangeably without effect in most cases, but the two units are not strictly identical.*

$$1.000000 \text{ milliliter} = 1.000028 \text{ cubic centimeter}$$

Since volume measurements cannot be made with an accuracy of 1 part in 36,000, no serious error results from interchanging the units.

Temperature Effect on Volume Measurements

The volume of a given mass of liquid varies with temperature. The container used to measure the volume also varies with temperature. Consequently highly accurate volumetric measurements require that the effect of temperature be taken into account.

In the United States, the standard temperature for glassware has been selected as 20°C. Volumetric glassware are constructed and calibrated to hold or to deliver the designated volumes at 20°C. At higher temperatures slightly larger volumes are involved and at lower temperatures slightly smaller volumes are involved. Only when extreme accuracy is required are temperature corrections for changes in volume of glass containers required. The cubical expansion of soft glass is about 0.000026 ml per degree for each ml of volume. For pyrex glass the expansion is only about 0.000011 ml per degree per ml. For a 50 ml pipet or buret the required correction corresponds to only 0.0013 ml for soft glass and 0.00055 ml for pyrex per degree. Obviously, since most laboratories are air conditioned, temperature changes would not cause a significant change in volume. The same statement does not always hold for large volumetric flasks. Regardless

---

*The twelfth "Conference Generale des Poids et Mesures" in 1964 decided to rescind the definition of liter and redefine it as a special name for the cubic decimeter making milliliter identical to cubic centimeter. The resolution further pointed out that *liter* should not be used to express results of high precision. There is little indication that analytical chemists have adopted the new definition. This book uses the classical definition as given above.

of the container size temperature corrections are meaningless if the container has not been calibrated. Permissible marking error in construction of commercial apparatus is greater than volume changes due to temperature changes.

The cubical expansion of water and dilute aqueous solutions is 0.00025 ml per degree for each ml. Standard solutions which are diluted to a given volume at one temperature do not have the same concentration at a different temperature. For highest accuracy, concentration or volume changes with temperature should be considered.

EXAMPLE 6-1: A solution was found to be $0.1010N$ at 20°C. The solution was used at 23°C in a titration, 42.12 ml being required to reach the end point. Calculate the volume which would have been required if the solution were used at 20°C (where its concentration is known).

The change in volume is given by

$$\Delta V = 0.00025 \text{ ml/deg ml} \times 3 \text{ deg} \times 42.12 \text{ ml}$$

$$= 0.03 \text{ ml}$$

The solution occupies a larger volume at 23°C than at 20°C, therefore, the 42.12 ml would have occupied only 42.09 ml at 20°C.

$$V = 42.12 - 0.03 = 42.09 \text{ ml}$$

The following example illustrates a unique situation in which both the expansion of the flask and of the solution are considered.

EXAMPLE 6-2: A Pyrex flask correctly marked to hold 1000.00 ml at 20°C was filled to the mark with a dilute aqueous solution at 20°C. The flask and its contents were warmed to 30°C. Calculate how high above the mark the meniscus now occurs. Assume the diameter of the flask to be 16 mm.

The volume of solution in the flask is given by

$$1000.00 \text{ ml} + 0.00025 \text{ ml/deg ml} \times 10 \text{ deg} \times 1000 \text{ ml}$$

$$= 1000.00 + 2.50$$

$$= 1002.50 \text{ ml}$$

The volume of the flask itself is given by

$1000.00 \text{ ml} + 0.000011 \text{ ml/deg ml} \times 10 \text{ deg} \times 1000 \text{ ml}$

$$= 1000.00 + 0.11$$

$$= 1000.11 \text{ ml}$$

The volume of solution above the mark is, therefore,

$$1002.50 - 1000.11 = 2.39 \text{ ml}$$

To determine the height above the mark, calculate what height of cylinder has a volume of 2.39 ml (2.39 cm^3).

$$V = \pi r^2 h$$

$$2.39 \text{ cm}^3 = 3.14 \times 0.8^2 \text{ cm}^2 \times h$$

$$h = \frac{2.39}{3.14 \times 0.8^2}$$

$$= 1.2 \text{ cm}$$

The meniscus is, therefore, 1.2 cm above the mark.

Standard Solutions

A standard solution is a solution having an exactly known concentration. It is normally used as a titrating reagent and when used as such is referred to as a titrant. The concentration of standard solutions is determined by one of three ways. One, a known weight of a *primary standard* chemical is diluted to an exact volume in a volumetric flask. Two, an approximately known concentration is prepared and the solution is titrated against a weighed amount of primary standard. This procedure is known as *standardization*. The third way to determine the strength of a standard solution is to compare it by titration with a solution previously standardized by one of the first two methods. This latter type of standard solution is known as a secondary standard. A standard solution prepared by the first or second method is preferable because any error in standardization procedure is compounded in the third method.

In order for a chemical to serve as a primary standard, several requirements must be met.

1. The material must be obtained in a high degree of purity. Although 100% purity is desirable, if an exact purity is known and the impurities are known not to interfere, the material can be used as a primary standard.
2. The material must react in a single stoichiometric reaction.
3. The material must be capable of being dried without undergoing a change in composition.
4. The material must be stable to its environment. It should not undergo oxidation nor other changes. In the dry form, the material should be stable for an indefinite period. In solution, the material must also be stable for as long as the solution is needed.
5. The material must neither gain nor lose water during weighing.
6. The error in the determination of the end point must be negligible or easy to determine experimentally.

In addition to the above necessary requirements the following features are highly desirable.

7. The material should have a high *equivalent weight* because the the higher the equivalent weight, the greater the weight of primary standard required for the titration. Therefore, the relative error in weighing will be lower.
8. The material should be readily attainable at a reasonable cost.

Equivalent Weights of Substances

In the fundamental principle of a titration, reference was made to the number of equivalent weights of reacting substances. Equivalent weights are used when concentrations are expressed in normality units. A $1N$ solution contains one equivalent per liter or one milliequivalent per milliliter. Normal solutions are such that 1 equivalent or milliequivalent of titrant will react with 1 equivalent or milliequivalent of substance titrated.

The determination of the equivalent weight of a substance depends upon the type of reaction the substance undergoes. The general equation for the equivalent weight is

$$\text{equivalent weight} = \frac{\text{molecular weight or formula weight}}{n}$$

Acid-Base Titrations

The equivalent weight of an acid is the molecular weight divided by the number of replaceable hydrogen atoms for the reaction it undergoes. The equivalent weight of a base is defined as the molecular weight divided by the number of hydrogen atoms required to neutralize the base.

$$eq\ wt\ =\ \frac{mol\ wt}{no.\ of\ H^+}$$

Some examples are given in Table 6-1.

Table 6-1

Equivalent Weights of Acids and Bases

Acid or base	n	Equivalent weight
HCl	1	36.46
H_2SO_4	2	49.04
H_3PO_4	1 or 2	98.00 or 49.00
NaOH	1	40.00
$Ba(OH)_2$	2	85.67
Na_2CO_3	1 or 2	106.00 or 53.00

In regards to the third entry in Table 6-1, for the reaction

$$H_3PO_4 + OH^- \rightarrow H_2PO_4^- + H_2O$$

the equivalent weight of H_3PO_4 is equal to the molecular weight. The end point for this reaction is indicated by the color change of methyl orange. For the reaction

$$H_3PO_4 + 2OH^- \rightarrow HPO_4^{--} + 2H_2O$$

the equivalent weight of H_3PO_4 is equal to one-half the molecular weight. This end point is indicated by the color change of phenol-phthalein.

In regards to the sixth entry in Table 6-1, for the reaction

$$Na_2CO_3 + H^+ \rightarrow NaHCO_3 + Na^+$$

the equivalent weight of Na_2CO_3 is equal to the molecular weight. The end point for this reaction is indicated by the color change of phenolphthalein. For the reaction

$$Na_2CO_3 + 2H^+ \rightarrow H_2CO_3 + 2Na^+$$

the equivalent weight of Na_2CO_3 is equal to one-half the molecular weight. This end point is indicated by the color change of methyl orange.

As indicated above, the equivalent weight of weak polybasic acids (and weak polyacidic bases) depends upon the reaction involved. For these solutions, molarity is best used to prevent ambiguity.

Redox Titrations

The equivalent weight of a substance undergoing oxidation or reduction is equal to the molecular weight divided by the *total* number of electrons gained or lost *per* molecule in the particular reaction.

$$\text{eq wt} = \frac{\text{mol wt}}{\text{no of } e^- \text{ gained or lost}}$$

The following half-reactions illustrate the determination of equivalent weight in redox reactions.

$$MnO_4^- + 8H^+ + 5e^- \rightarrow Mn^{++} + 4H_2O$$

$$\text{eq wt } KMnO_4 = \frac{158.04}{5} = 31.61$$

$$MnO_4^- + 4H^+ + 3e^- \rightarrow MnO_2 + 2H_2O$$

$$\text{eq wt } KMnO_4 = \frac{158.04}{3} = 52.68$$

$$Fe^{+++} + e^- \rightarrow Fe^{++}$$

$$\text{eq wt } Fe = 58.85$$

$$Sn^{+4} + 2e^- \rightarrow Sn^{++}$$

$$\text{eq wt } Sn = \frac{118.69}{2} = 59.34$$

$$Cr_2O_7^{--} + 14H^+ + 6e^- \rightarrow 2Cr^{+++} + 7H_2O$$

$$\text{eq wt } K_2Cr_2O_7 = \frac{294.19}{6} = 49.03$$

Precipitation Titrations

The equivalent weight of a cation in a titration reaction is the formula weight of the cation divided by the charge on the ion. The equivalent weight of the anion in such a reaction is the formula weight of the anion divided by the number of metal ion equivalents with which it reacts. The following reactions serve as examples:

$$Ag^+ + Cl^- \rightarrow AgCl$$

$$\text{eq wt } Ag^+ = 107.88$$

$$\text{eq wt } Cl^- = 35.45$$

$$2Ag^+ + CrO_4^{--} \rightarrow Ag_2CrO_4$$

$$\text{eq wt } Ag^+ = 107.88$$

$$\text{eq wt } CrO_4^{--} = \frac{115.99}{2} = 58.00$$

Complex-formation Titrations

The equivalent weight of cations and anions undergoing complex-formation reactions is determined as above for precipitation reactions. For example,

$$Cu^{++} + 4CN^- \rightleftharpoons Cu(CN)_4^{--}$$

$$\text{eq wt } Cu^{++} = \frac{63.54}{2} = 31.77$$

$$\text{eq wt } CN^- = \frac{26.02}{1/2} = 52.04$$

It should be clear from the preceding discussion that the equivalent weight of a substance is not a constant but depends upon the reaction in which the standard solution is used.

Whenever the possibility of ambiguity arises concerning equivalent weight it is advisable either to use molarity or to express the concentration in both terms. For example, a $1N$ ferric chloride solution may be $1M$:

$$Fe^{+++} + e^- \rightarrow Fe^{++}$$

or $(1/3)M$

$$Fe^{+++} + 3OH^- \rightarrow Fe(OH)_3$$

In this case the concentrations should be labeled

$$1N\, FeCl_3 \left(\frac{M}{1}\right) \quad \text{or} \quad 1N\, FeCl_3 \left(\frac{M}{3}\right)$$

Calculations in Volumetric Analysis

There are basically four types of calculations upon which volumetric analysis is based. The first three correspond to the three methods of determining the concentration of standard solutions. The fourth is used for calculating the percentage composition from titration data.

The following relationships based upon the molarity system are used in volumetric analysis:

$$M = \frac{\text{moles}}{\text{liter}}$$

$$M = \frac{\text{millimoles}}{\text{milliliter}} = \frac{\text{mmoles}}{\text{ml}}$$

$$V \text{ (liters)} \times M = \text{moles}$$

$$V \text{ (ml)} \times M = \text{mmoles}$$

$$\text{moles} \times \text{molecular wt} = \text{wt in grams}$$

$$\text{mmoles} \times \text{millimolecular wt} = \text{wt in grams}$$

Likewise, the relationships based upon the normality system are

$$N = \frac{\text{equivalents}}{\text{liter}}$$

$$N = \frac{\text{milliequivalents}}{\text{milliliter}} = \frac{\text{meqs}}{\text{ml}}$$

$$V \text{ (liters)} \times N = \text{equivalents}$$

$$V \text{ (ml)} \times N = \text{meqs}$$

$$\text{equivalents} \times \text{equivalent wt} = \text{wt in grams}$$

$$\text{meqs} \times \text{meq wt} = \text{wt in grams}$$

Since measurements in volumetric analysis are normally taken in milliliters, millimolecular weight and milliequivalent weight are much more useful units than molecular weight and equivalent weight because it eliminates the shifting of the decimal point for buret readings.

Dilution of a Known Weight of Primary Standard to a Known Volume

If a substance meets the requirements of a primary standard, a standard solution can be prepared by accurate dilution of a known weight of the primary standard. This is the most accurate method for preparing a standard solution. The general equation used to calculate the normality of such a solution is

$$N = \frac{\text{wt of primary standard}}{\text{meq wt of primary standard} \times V_{\text{ml (dilution)}}} \tag{6-1}$$

Examination of Equation (6-1) shows why it is valid:

Obviously, the number of milliequivalents of the primary standard weighed must be equal to the number of milliequivalents of the primary standard in the solution

$$\text{meqs weighed} = \text{meqs in solution}$$

The number of milliequivalents weighed is given by Equation (6-2).

$$\text{meqs weighed} = \frac{\text{wt of primary standard}}{\text{meq wt of primary standard}} \tag{6-2}$$

The number of milliequivalents of primary standard in the solution is given by Equation (6-3).

$$\text{meqs in solution} = V_{\text{ml(dilution)}} \times N \qquad \textbf{(6-3)}$$

Equations (6-2) and (6-3) can be combined to give Equation (6-1).

$$\frac{\text{wt of primary standard}}{\text{meq wt of primary standard}} = V_{\text{ml(dilution)}} \times N$$

and

$$N = \frac{\text{wt of primary standard}}{\text{meq wt of primary standard} \times V_{\text{ml(dilution)}}} \qquad \textbf{(6-1)}$$

EXAMPLE 6–3: A standard solution of $K_2Cr_2O_7$ is prepared by dissolving 2.4515 g of dried $K_2Cr_2O_7$ in water and dilution to 500 ml in a volumetric flask. Calculate the normality of the solution as a standard oxidizing agent. The milliequivalent weight of $K_2Cr_2O_7$ as an oxidant is 1/6000 of its molecular weight (0.04903 g).

$$N = \frac{2.4515 \text{ g}}{0.04903 \text{ g/meq} \times 500 \text{ ml}}$$

$$= 0.1000N$$

Standardization vs. a Primary Standard

If a standard solution is required of a substance which does not meet primary standard requirements, it can be standardized by titration with a known amount of primary standard. The approximate amount of reagent is diluted to give a concentration approximately equal to the desired concentration. After the solution is thoroughly mixed, a portion of it is placed in a buret and it is titrated against a weighed amount of primary standard. The normality of the solution is given by

$$N = \frac{\text{wt of primary standard}}{\text{meq wt of primary standard} \times V_{\text{ml(titration)}}} \qquad \textbf{(6-4)}$$

This equation is explained as follows:

The fundamental principle of volumetric analysis is that at the equivalence point in a titration the number of milliequivalents of the two reacting species are equal. Therefore, meqs of primary standard = meqs of standard solution

The number of milliequivalents of primary standard is given by Equation (6-5).

$$\text{meqs of primary standard} = \frac{\text{wt of primary standard}}{\text{meq wt of primary standard}} \quad \text{(6-5)}$$

The number of milliequivalents of standard solution is given by Equation (6-6).

$$\text{meqs of standard solution} = V_{ml(\text{titration})} \times N \quad \text{(6-6)}$$

Equations (6-5) and (6-6) can be combined to give Equation (6-4).

$$\frac{\text{wt of primary standard}}{\text{meq wt of primary standard}} = V_{ml(\text{titration})} \times N$$

and

$$N = \frac{\text{wt of primary standard}}{\text{meq wt of primary standard} \times V_{ml(\text{titration})}} \quad \text{(6-4)}$$

EXAMPLE 6-4: A solution of hydrochloric acid is standardized by titration against pure sodium carbonate. A 0.2573 g sample of sodium carbonate requires 42.12 ml of the HCl for neutralization to the methyl orange end point. Calculate the normality of the HCl.

$$Na_2CO_3 + 2HCl \rightarrow 2NaCl + H_2CO_3$$
$$ \llcorner\!\!\rightarrow H_2O + CO_2$$

$$N = \frac{0.2573 \text{ g}}{\frac{105.99}{2000} \text{ g/meq} \times 42.12 \text{ ml}}$$

$$= 0.1153 N$$

Standardization vs. a Standardized Solution

The third method for preparing a standard solution is to ~~titrate a sample of the solution against a known volume of another standard solu~~-

tion. Alternatively, the standard solution can be titrated against a
known volume of the solution to be standardized. A solution standardized in this manner is not as accurate as one prepared by the first
two methods because it includes the experimental error involved in two
standardizations. Solutions are standardized in this manner when a
primary standard is not available or for expedience. For example,
sodium hydroxide is often standardized with potassium acid phthalate
and then used to standardize hydrochloric acid, thus eliminating the
drying and weighing of a second primary standard. Equation (6-7) is
used to calculate the normality of a solution prepared in this manner.

$$N_2 = \frac{V_1 \times N_1}{V_2} \qquad (6\text{-}7)$$

where V_1 and N_1 are the volume and normality of the standard solution and V_2 and N_2 are the same for the solution being standardized.

Equation (6-7) also follows directly from the fundamental principle
of titrations.

meqs of solution 1 = meqs of solution 2

meqs of solution 1 = $V_1 \times N_1$

meqs of solution 2 = $V_2 \times N_2$

therefore

$$V_1 \times N_1 = V_2 \times N_2$$

and

$$N_2 = \frac{V_1 \times N_1}{V_2} \qquad (6\text{-}7)$$

EXAMPLE 6-5: A hydrochloric acid solution was standardized
with standard sodium hydroxide solution. A 50.00 ml sample of
the HCl solution required 44.35 ml of 0.1102N NaOH for neutralization. Calculate the normality of the HCl.

$$N = \frac{44.35 \text{ ml} \times 0.1102N}{50.00 \text{ ml}}$$

$$= 0.0977N$$

Preparation of Standard Solutions
from Concentrated Reagents

~~To prepare a standard solution from a concentrated laboratory reagent (e.g.,~~ HCl, H_2SO_4, NH_3) ~~the specific gravity and percentage composition of the reagent is required.~~ A solution having the approximate required normality is prepared and the solution is standardized by either the second or third method listed previously.

The calculation of the approximate volume of concentrated reagent required to prepare a solution of approximately a certain normality is illustrated by the following example.

> **EXAMPLE 6-6:** Calculate the volume of concentrated sulfuric acid having a specific gravity of 1.842 and containing 96.0% H_2SO_4 by weight required to prepare 2.0 liters of $0.20N$ H_2SO_4.
>
> The weight of pure hydrogen sulfate required is given by Equation (6-8).

$$\text{weight } H_2SO_4 \text{ needed} = 2.0 \text{ l} \times 0.2 \text{ eq/l} \times \frac{98.08}{2} \text{ g/eq} \quad \textbf{(6-8)}$$

The volume of concentrated H_2SO_4 required to give the above weight is calculated as follows: Let V equal volume of H_2SO_4 needed, then the weight of H_2SO_4 in V is the following:

$$\text{weight} = V \text{ ml} \times 1.842 \text{ g/ml} \times 0.96 \quad \textbf{(6-9)}$$

Equations (6-8) and (6-9) can be combined to give Equation (6-10).

$$V \times 1.842 \times 0.96 = 2.0 \times 0.20 \times 49.04 \quad \textbf{(6-10)}$$

$$V = \frac{2.0 \times 0.20 \times 49.04}{1.842 \times 0.96}$$

$$= 11.1 \text{ ml}$$

Preparation of Exact Predetermined Concentration

Occasionally it is desired to prepare a standard solution of a certain exact concentration. This is easily done if the reagent is a primary standard. The exact amount required is accurately weighed and is diluted to a known volume in a volumetric flask. When the substance is not a primary standard direct weighing cannot give a known concentration. In this case a solution slightly stronger than the one desired is prepared, the solution is standardized and accurately diluted to the desired value. Example 6-7 will serve to illustrate this technique.

EXAMPLE 6-7: Prepare one liter of exactly $0.2000N$ H_2SO_4.
One should prepare more than 1000 ml since some will be consumed in the standardization. Also, the concentration should be made slightly stronger than $0.2000N$. We will assume that 1250 ml of $0.2100N$ will be needed.

$$V = \frac{1250 \times 0.2100 \times 0.04904}{1.842 \times 0.96}$$

$$= 7.3 \text{ ml}$$

Approximately 1250 ml of water measured with a graduate is placed in a bottle and 7.3 ml of concentrated sulfuric acid is added with a Mohr pipet. The solution is thoroughly mixed and the solution is standardized. We will assume that the standardization showed the solution to be $0.2086N$. The volume of $0.2086N$ acid required to give one liter of $0.2000N$ acid is given by

$$V = \frac{1000 \times 0.2000}{0.2086}$$

$$= 958.8 \text{ ml}$$

The accurate measurement of 958.8 ml is not feasible so the dilution is done as follows. Measure 41.2 ml of water from a buret into a dry liter volumetric flask. Then fill the flask to the mark with the $0.2086N$ acid. After mixing, the solution should be checked by titration against a standard base.

Calculation of Percentage from Titrations with Standard Solutions

The fourth basic type of calculation of volumetric analysis is used to calculate the results of a determination. Equation (6-11) is the general formula used.

$$\% Y = \frac{V \times N \times \text{meq wt of } Y \times 100}{\text{sample wt}} \qquad (6\text{-}11)$$

The explanation of Equation (6-11) follows:

The volume in milliliters times the normality yields the number of milliequivalents of the titrant.

$$V_{\text{ml}} \times N = \text{meqs of titrant} \qquad (6\text{-}12)$$

The fundamental principle of titrations tells us that the number of milliequivalents of titrant is equal to the number of milliequivalents of substance Y titrated. Therefore, Equation (6-12) also yields the number of milliequivalents of Y. The number of milliequivalents of Y times the milliequivalent weight of Y yields the weight of Y in grams.

$$V \times N \times \text{meq wt of } Y = \text{g of } Y \qquad (6\text{-}13)$$

The percentage of Y follows directly from Equations (6-13) and (6-14).

$$\% Y = \frac{\text{g of } Y}{\text{sample wt}} \times 100 \qquad (6\text{-}14)$$

Thus,

$$\% Y = \frac{V \times N \times \text{meq wt of } Y \times 100}{\text{sample wt}} \qquad (6\text{-}11)$$

EXAMPLE 6–8: A 1.4260 g sample of impure potassium acid phthalate (KHP) was dissolved and titrated with $0.1064N$ NaOH requiring 40.43 ml to reach the phenolphthalein end point. Calculate the percent KHP in the sample.

$$\% \text{KHP} = \frac{V_{\text{NaOH}} \times N_{\text{NaOH}} \times \text{meq wt of KHP} \times 100}{\text{sample wt}}$$

$$= \frac{40.43 \times 0.1064 \times 0.20423 \times 100}{1.4260}$$

$$= 61.61\% \text{ KHP}$$

Back-titrations

The requirements for a successful titration are (1) a single stoichiometric reaction, (2) a complete and rapid reaction, and (3) an end point that is detectable. If the reaction is not rapid, the principle of back-titration may make the reaction suitable for a volumetric determination.

In a back-titration, an excess amount of titrant (reagent A) is added and the amount of excess A is determined by a titration with a second standard solution (reagent B). The net number of milliequivalents of A reacting with the component being analyzed is equal to the total number of milliequivalents of A added minus the number of milliequivalents of B used in the back-titration.

$$\text{net meqs of } A = V_A N_A - V_B N_B \qquad (6\text{-}15)$$

The technique of back-titration is illustrated in the analysis of calcium carbonate in limestone. Calcium carbonate is insoluble in water but reacts with standard hydrochloric acid solutions. However, as the equivalence point is approached the rate at which the reaction proceeds becomes slower and slower. To titrate a limestone sample directly with no end-point error would require drop by drop addition of the acid, and then waiting until each drop had reacted before adding the next drop. This would require a prohibitively long time. In practice, a volume of acid 10%–20% in excess of the required amount is added. The excess acid acts as a driving force causing the limestone to completely react in a reasonable amount of time. The excess hydrochloric acid is determined by titration with standard sodium hydroxide.

EXAMPLE 6-9: A sample of limestone is analyzed as follows: A 0.2516 g sample is finely ground and suspended in 100 ml of water, 50.00 ml of $0.1128N$ HCl is pipetted into the flask carefully to prevent loss of sample due to effervescence. After the reaction is complete, the solution is boiled briefly to remove CO_2 and the excess acid is neutralized with 6.22 ml of $0.1022N$ NaOH. Calculate the percent $CaCO_3$ in the sample.

$$\%CaCO_3 = \frac{(V_{HCl} \times N_{HCl} - V_{NaOH} \times N_{NaOH}) \times \dfrac{CaCO_3}{2000} \times 100}{\text{sample wt}}$$

$$= \frac{(50.00 \times 0.1128 - 6.22 \times 0.1022) \times 0.05004 \times 100}{0.2516}$$

$$= 99.44\% \, CaCO_3$$

Sometimes the detection of the end point in a titration is sharper when approached from one side than from the other. This is often true in EDTA titrations which are performed by a back-titration.

EXAMPLE 6-10: A sample containing soluble iron(III) is analyzed as follows: A 0.4592 g sample is dissolved, 25.00 ml of 0.0500M EDTA is added, and the excess EDTA is titrated with 2.86 ml of 0.0210M copper(II). Both metals react with EDTA on a 1:1 mole basis. Calculate the percent Fe(III) in the sample.

$$\%Fe(III) = \frac{(25.00 \times 0.0500 - 2.86 \times 0.0210) \times 0.05585 \times 100}{0.4592}$$

$$= 14.47\%$$

Indirect Titrations

In an indirect titration, the substance being measured does not react directly with the titrant. The titrant reacts with a species stoichiometrically equivalent to the substance being measured. The determination of calcium in limestone by titration with potassium permanganate is an example of an indirect titration.

The weighed limestone sample is dissolved in hydrochloric acid, the calcium ion is precipitated as calcium oxalate which is washed, filtered, and dissolved in sulfuric acid which converts the oxalate ion into oxalic acid. The oxalic acid is then titrated with standard permanganate solution. The titration determines the milliequivalents of oxalate ion which is equal to the milliequivalents of calcium in the sample.

EXAMPLE 6-11: A 0.2563 g sample of limestone was analyzed for calcium by the method described above. The titration required 41.92 ml of 0.1088N KMnO$_4$. Calculate the percent calcium in the sample.

$$\%Ca = \frac{V_{KMnO_4} \times N_{KMnO_4} \times meq \, wt \, Ca \times 100}{sample \, wt}$$

The equivalent weight of calcium must be determined. This is an oxidation-reduction titration, therefore, ~~the equivalent weight should be the atomic weight divided by the number of electrons gained or lost per atom of calcium.~~ Calcium does not undergo oxidation or reduction, so no electrons are gained or lost. Each mole of calcium ions is equivalent to one mole of oxalate ions. The oxalate ion does undergo oxidation, losing two electrons per ion. The equivalent weight of oxalate ion is therefore one-half its formula weight. Since calcium and oxalate ions react on a $1:1$ mole basis, the equivalent weight of calcium is one-half its atomic weight.

$$\%\text{Ca} = \frac{41.92 \times 0.1088 \times \dfrac{40.08}{2000} \times 100}{0.2563}$$

$$= 35.66\% \text{ Ca}$$

Reported as percent CaO the results would be

$$35.66\% \times \frac{\text{CaO}}{\text{Ca}} = 49.90\% \text{ CaO}$$

Reported as percent $CaCO_3$ the results would be

$$35.66\% \times \frac{\text{CaCO}_3}{\text{Ca}} = 89.05\% \text{ CaCO}_3$$

Simultaneous Determination of Two Components in a Pure Mixture

The determination of the composition of a pure mixture of two substances both of which react with a titrant can be made with a single titration. The calculations are similar to those employed in the simultaneous gravimetric determination of two components of a mixture (see page 84).

EXAMPLE 6-12: A mixture consists of lithium carbonate and barium carbonate. A 0.5206 g sample is suspended in water, 50.00 ml of $0.2122N$ HCl is added, and the excess HCl is neutralized with 5.20 ml of $0.1028N$ NaOH. Calculate the percentage composition of the mixture.

$$\text{total meqs of carbonate} = \text{meqs HCl} - \text{meqs NaOH}$$

$$= 50.00 \times 0.2122 - 5.20 \times 0.1028$$

$$= 10.08$$

$$\text{meqs of Li}_2\text{CO}_3 + \text{meqs of BaCO}_3 = 10.08$$

Let X = weight of Li_2CO_3 in the sample, then

$$0.5206 - X = \text{weight of BaCO}_3 \text{ in the sample}$$

$$\text{meqs of Li}_2\text{CO}_3 = \frac{X}{\dfrac{Li_2CO_3}{2000}}$$

$$\text{meqs of BaCO}_3 = \frac{0.5206 - X}{\dfrac{BaCO_3}{2000}}$$

$$\frac{X}{0.03694} + \frac{0.5206 - X}{0.09868} = 10.08$$

$$X = 0.284 \text{ g}$$

$$\% \text{Li}_2\text{CO}_3 = \frac{0.284}{0.5206} \times 100 = 54.5\%$$

$$\% \text{BaCO}_3 = 45.5\%$$

It should be noted that one significant figure has been lost in the calculation. Simultaneous determinations do not yield results as accurate as would individual determinations of the two species. However, separation of the two components to allow individual analysis is not always feasible.

Another example of the determination of percentage composition of a mixture is in the analysis of fuming sulfuric acid samples. Fuming sulfuric acid is known as an oleum and consists of a solution of sulfur trioxide, SO_3, dissolved in anhydrous hydrogen sulfate, H_2SO_4. The sample is analyzed by dissolving a known weight of sample in water and titrating with standard base.

EXAMPLE 6-13. A sample of fuming sulfuric acid weighing 1.025 g is dissolved in water and titrated with 42.91 ml of $0.5022N$ NaOH solution. Calculate the percentage composition of the sample.

There are two methods which can be used to solve the problem. Since fuming sulfuric acid consists of a mixture of two pure components H_2SO_4 and SO_3, both of which react with NaOH, the problem can be solved as in the preceding example.

Let X = weight of H_2SO_4.

$$1.025 - X = \text{weight of } SO_3$$

$$\frac{X}{\frac{H_2SO_4}{2000}} + \frac{1.025 - X}{\frac{SO_3}{2000}} = 42.91 \text{ ml} \times 0.5022N$$

$$\frac{X}{.04904} + \frac{1.025 - X}{0.04003} = 21.55$$

$$X = 0.884 \text{ g}$$

$$\% H_2SO_4 = \frac{0.884}{1.025} \times 100 = 86.2\%$$

$$\% SO_3 = 13.8\%$$

The problem can also be solved on the basis of the sulfuric acid equivalency in the sample.

When the oleum is dissolved in water, the SO_3 combines with water to form H_2SO_4. The total percentage of acid expressed in terms of H_2SO_4 is given by

$$\frac{42.91 \text{ ml} \times 0.5022 \times \frac{H_2SO_4}{2000} \times 100}{1.025} = 103.1\%$$

Since the $H_2SO_4 + SO_3$ in the original mixture equals 100.0%, the 3.1% represents the water which combines with the SO_3 to form H_2SO_4. The percent SO_3 is given as follows:

$$3.1 \times \frac{SO_3}{H_2O} = \% SO_3$$

$$3.1 \times \frac{80.06}{18.02} = 13.8\% \ SO_3$$

$$100 - 13.8 = 86.2\% \ H_2SO_4$$

Aliquot Factors

The precision and accuracy in a volumetric analysis can sometimes be increased by weighing a large sample, diluting to a known volume, and then titrating a known fraction of that volume.

EXAMPLE 6-14: A 5.0000 g sample of soda ash was dissolved in water and diluted to 500.0 ml. A 50.00 ml portion was titrated with 0.1800N HCl requiring 46.25 ml for neutralization. Calculate the percentage of sodium carbonate in the sample.

Only one-tenth of the sample is titrated so the sample weight must be corrected to reflect this amount of sample. The percentage of sodium carbonate is given as follows:

$$\% Na_2CO_3 = \frac{46.25 \times 0.1800 \times \dfrac{Na_2CO_3}{2000} \times 100}{5.0000 \times \dfrac{50.00}{500.0}}$$

$$= 88.25\% \ Na_2CO_3$$

The aliquot factor 500.0/50.00 or 10 is usually placed in the numerator. Since only one-tenth of the sample weight is titrated, titration of the total sample would require ten times as much titrant. The general formula to use when an aliquot portion of the sample is titrated is

$$\% Y = \frac{V_T \times N_T \times \text{meq wt of } Y \times F \times 100}{\text{sample wt}}$$

where V_T and N_T are the volume in ml and the normality of the titrant, F is the aliquot factor, and the sample weight is the total weighed sample.

The aliquot factor technique is used whenever the weighing of individual samples might yield erroneous results. This could be the case if the sample was not completely homogeneous. The technique can also be useful when the sample is slightly hygroscopic since weighing a single large sample would be quicker and easier than replicate smaller samples.

Factor Weight

The calculation of percentage can be simplified if the sample weight is adjusted so that the volume of titrant will have a simple relationship to the percentage. The weight is usually such that the volume in milliliters is equal to the percentage or some simple ratio of the percentage.

> **EXAMPLE 6-15:** Calculate the sample weight to use in the titration of chloride with $0.1000N$ $AgNO_3$ such that the volume of $AgNO_3$ is one-half the percentage of chloride.
> Set up the equation for the percentage.

$$\%Cl = \frac{V \times N \times \frac{Cl}{1000} \times 100}{\text{sample wt}} \qquad (6\text{-}16)$$

Set up the equation for the relationship between the volume and the percentage.

$$V = 0.5 \times \%Cl \qquad (6\text{-}17)$$

Substitute Equation (6-17) into Equation (6-16).

$$\%Cl = \frac{0.5 \times \%Cl \times N \times \frac{Cl}{1000} \times 100}{\text{sample wt}}$$

$$\text{sample wt} = \frac{0.5 \times \%Cl \times 0.1000 \times 0.03545 \times 100}{\%Cl}$$

$$= 0.1773 \text{ g}$$

A slight adaption of this method is often used in industrial control laboratories for routine assay analysis. Large numbers of titrations are necessary and the percentage composition is usually quite similar. Consider the analysis of aspirin samples for acetyl salicylic acid (ASA). The weighed aspirin sample is dissolved in water and titrated with standard base using phenolphthalein as an indicator. A convenient sample size (e.g., 1.000 g) is weighed to the nearest milligram and the sample is titrated with a standard base (e.g., $0.2000N$). A table is prepared giving the percentage composition for various titration volumes. The vicinity of the end point can be approached rapidly and the exact end point detected without much waste of time. Errors in calculations are avoided in the analysis when values from the table are used. Table 6-2 shows such a table for percentage purity of ASA samples.

Table 6-2

Percentage of Acetyl Salicylic Acid in Aspirin
(1.000 g sample, 0.2000N NaOH)

V_{NaOH}	%ASA
27.80	100.2
27.75	100.0
27.70	99.8
27.65	99.7
27.60	99.5
27.55	99.3
27.50	99.1

Questions

1. State the fundamental principle of titrations and show how it can be applied to the calculations of the result of a titration.

2. List the necessary criteria which must be met in order for a reaction to be suitable for a volumetric determination.

3. Discuss the meaning and cause of the end point error in a titration.

4. Volumetric methods of analysis can be categorized into four major types. List the types and explain how the equivalent weight of a substance undergoing each major type of analysis is calculated.

5. Why is it necessary to have a standard temperature for volumetric analysis?

6. Comment on the validity of the statement: "A liter is defined as 1000.0 g mass of pure water at 3.98°C and 1 atmosphere."

7. List the necessary and desirable characteristics of a primary standard chemical.

8. A standard solution can be prepared by three separate methods. Describe each method.

9. What is a secondary standard solution?

10. List two advantageous applications of the technique of back-titration.

11. List two advantageous applications of the technique of aliquot factors.

12. What is a factor weight?

Problems

1. What fraction of a molecular weight is the milliequivalent weight of the following substances?
   (a) KOH   (b) H_2SO_4 (c) $Ca(OH)_2$
   (d) H_3PO_4 in a reaction in which it is converted into Na_2HPO_4.

2. A standard solution was prepared and standardized at 25°C. If this solution is used in a titration at 28°C, calculate the relative percent error introduced by failure to apply the appropriate temperature correction. Would this omission cause a positive or a negative error in the percent composition as calculated from the titration?

3. Calculate the volume of concentrated HCl having a density of 1.188 g/ml and containing 38% HCl by weight needed to prepare 2 liters of approximately $0.20N$ hydrochloric acid.

4. Calculate the volume of distilled water which must be added to 1000.0 ml of $0.1062N$ NaOH to give a solution having a normality of exactly 0.1000.

5. Calculate the normality of a sodium chloride solution prepared by dissolving 2.9216 g NaCl in water and dilution to 500.0 ml.

6. Calculate the normality of a silver nitrate solution, $AgNO_3$, prepared by dissolving 17.856 g of $AgNO_3$ in water and dilution to 1000.0 ml.

7. A 0.8056 g sample of primary standard potassium acid phthalate, $KHC_8H_4O_4$, was dissolved in water and titrated with 41.22 ml of sodium hydroxide. Calculate the normality of the base.

8. A 25.00 ml sample of HCl is neutralized by the addition of 32.16 ml of $0.1064N$ NaOH. Calculate the normality of the HCl solution.

9. A 25.00 ml sample of H_2SO_4 is neutralized by the addition of 17.46 ml of $0.2054N$ NaOH. Calculate the normality and the molarity of the H_2SO_4 solution.

10. A 1.2908 g sample containing impure potassium acid phthalate was titrated with 29.68 ml of $0.1302N$ NaOH. Calculate the percentage of KHP in the sample.

11. A 1.0000 g sample of impure sodium carbonate, Na_2CO_3, was dissolved in water and titrated according to the following reaction with 42.40 ml of hydrochloric acid, 25.00 ml of which was required to neutralize 48.21 ml of $0.1022N$ NaOH. Calculate the percentage of Na_2CO_3 in the sample.

$$Na_2CO_3 + 2HCl \rightarrow 2NaCl + H_2O + CO_2$$

$$N = \frac{48.21 \ (0.1022)}{25}$$

12. A 0.2536 g sample of pure sodium carbonate was dissolved in water and titrated with 31.25 ml of hydrochloric acid to the methyl orange end point. Calculate the normality of the acid solution.

13. A 0.2560 g sample of a pure monobasic acid is dissolved in water and titrated with 29.72 ml of 0.1022N NaOH. Calculate the molecular weight of the acid.

14. A 0.3025 g sample of impure $CaCO_3$ was dissolved in 50.00 ml of 0.1422N HCl according to the following reaction. The excess acid was neutralized with 10.64 ml of 0.1080N NaOH. Calculate the percentage of $CaCO_3$ in the sample.

$$CaCO_3 + 2HCl \rightarrow CaCl_2 + CO_2 + H_2O$$

15. A 0.5020 g sample of an organic compound was analyzed by the Kjeldahl method. After the sample was digested and the nitrogen converted into ammonia, the ammonia was distilled into a flask containing 25.00 ml of 0.1000N HCl. The excess HCl required 3.12 ml of 0.1104N NaOH for neutralization. Calculate the percentage of N in the sample.

16. Calculate the weight of sample which should be taken so that the milliliters of 0.1000N AgNO$_3$ when multiplied by 2 will equal the percentage of chloride in the sample.

17. A 0.3250 g sample containing only NaCl and KCl requires 38.50 ml of 0.1380N AgNO$_3$ to precipitate all the chloride. Calculate the percentage of NaCl in the sample.

18. A 0.5500 g sample of pure $CaCO_3$ mixed with pure $SrCO_3$ requires 28.76 ml of 0.3120N sulfuric acid for neutralization. Calculate the weight of $SrCO_3$ in the sample.

19. A 5.3260 g sample of soda ash was dissolved in water and diluted to 500.0 ml. A 50.0 ml aliquot required 48.22 ml of 0.2031N HCl for neutralization. Calculate the percentage of Na_2CO_3 in the sample.

20. A mixture of pure acetic acid and acetic anhydride is dissolved in water and titrated with NaOH. The acidity of the sample expressed in terms of $HC_2H_3O_2$ is 116.0%. Calculate the composition of the original mixture (acetic anhydride reacts with water to form acetic acid $(CH_3CO)_2O + H_2O \rightarrow 2HC_2H_3O_2$).

21. A 1.005 g sample of fuming sulfuric acid was carefully dissolved in water and 44.62 ml of 0.5012N NaOH was required to neutralize the acid. Calculate the percentage of SO_3 dissolved in anhydrous H_2SO_4 to prepare the oleum.

7

Calculations Involving Solutions of Acids and Bases

Acid-Base Theories

Several acid-base theories have been proposed to explain the acidic and basic properties of substances. The earliest and simplest classification is the *Arrhenius theory,* in which an acid is any substance which ionizes in water to give hydrogen ions H^+ or hydronium ions H_3O^+. A hydronium ion is simply a hydrated hydrogen ion or hydrated proton. The proton cannot exist unhydrated; indeed probably no ions exist unhydrated in an aqueous solution. Furthermore, the hydrated proton does not exist as H_3O^+ but rather in a more highly hydrated form. Studies have indicated the species probably to be $H_9O_4^+$ or $H^+ \cdot 4H_2O$.* Consequently in this text, it will be deemed as accurate to write H^+ for the species as it is to write H_3O^+. An Arrhenius base is a substance which ionizes in water to give hydroxyl ions OH^-.

In 1923, Brønsted, and in 1924, Lowry, working independently, proposed a new system now known as the *Brønsted-Lowry theory.* This

---

*H.L. Cleaver, *J. Chem. Ed.,* **40**, 637 (1963).

theory broadened the Arrhenius concept. A Brønsted-Lowry acid is
any substance which can give up a proton, and a Brønsted-Lowry base
is any substance which can accept a proton. When an acid gives up a
proton, the remaining species has a certain proton affinity and hence
is a base. This base is known as the conjugate base of the acid, and the
two forms are known as an acid-base pair.

$$acid \rightleftharpoons H^+ + base$$

Specific examples of conjugate acid-base pairs are the following:

$$HCl \rightleftharpoons H^+ + Cl^- \qquad\qquad (7\text{-}1)$$

$$HOAc \rightleftharpoons H^+ + OAc^- \qquad\qquad (7\text{-}2)$$

$$H_3O^+ \rightleftharpoons H^+ + H_2O \qquad\qquad (7\text{-}3)$$

$$H_2O \rightleftharpoons H^+ + OH^- \qquad\qquad (7\text{-}4)$$

The chloride and acetate ions are conjugate bases of hydrochloric acid
and acetic acid respectively. You will further note that in Equation
(7-3), water is a conjugate base of the hydronium ion, whereas in Equation (7-4), water is the conjugate acid of the base hydroxide ion. In
order for a Brønsted-Lowry acid to give up its proton, a proton ac-
ceptor must be available since its high charge density precludes a proton's independent existence in solution. This leads to the general
equation

$$acid_1 + base_2 \rightleftharpoons acid_2 + base_1 \qquad\qquad (7\text{-}5)$$

Often the solvent itself serves as one of the two conjugate acid-base
pairs as is illustrated in Equations (7-6) and (7-7).

$$HOAc + H_2O \rightleftharpoons H_3O^+ + OAc^- \qquad\qquad (7\text{-}6)$$

$$H_2O + NH_3 \rightleftharpoons NH_4^+ + OH^- \qquad\qquad (7\text{-}7)$$

The solvent water acts as a base in Equation (7-6) and as an acid in
Equation (7-7). Equation (7-8) represents the reaction between a strong
acid and a strong base.

$$H_3O^+ + OH^- \rightleftharpoons H_2O + H_2O \qquad\qquad (7\text{-}8)$$

The extent to which any of the above equilibrium reactions proceeds towards completion is controlled by the relative ease of transferring a proton from $acid_1$ to $base_2$ compared to a proton transfer from $acid_2$ to $base_1$. For example, Equation (7-8) representing the reaction between hydrochloric acid and sodium hydroxide proceeds nearly to completion because the hydroxide ion ($base_2$) has a stronger affinity for protons than does water ($base_1$). On the other hand, the affinity of the acetate ion ($base_1$) is not as strong as OH^-, but is stronger than the affinity of H_2O ($base_2$) for protons. Therefore, reaction (7-6) will only proceed slightly from left to right before reaching equilibrium.

It is obvious from the preceding that the tendency for an acid to lose its proton is related to the affinity of the solvent for the proton. The same is true for the tendency of a base to accept a proton. Therefore, the solvent is extremely important in determining the apparent strength of an acid.

Dissociation of Water and Ion Product of Water

Pure water is slightly ionized into hydrogen ions and hydroxyl ions. Although this degree of ionization is extremely small, it is of extreme importance in the calculations involving conditions of equilibrium in chemical processes.

The dissociation of water is a reversible reaction and can be expressed by Equation (7-9).

$$H_2O \rightleftharpoons H^+ + OH^- \tag{7-9}$$

By application of the law of mass action, the reaction involved in Equation (7-9) can be expressed as shown in Equation (7-10).

$$K = \frac{[H^+][OH^-]}{[H_2O]} \tag{7-10}$$

or

$$K[H_2O] = [H^+][OH^-]$$

Equation (7-10) can be simplified by the assumption that the degree of ionization does not affect the water concentration. This follows from the fact that a liter of water contains approximately 55.5 moles of water and the degree of dissociation is quite low. Since both K and $[H_2O]$ are constant, they can be combined into a single constant and this single constant is known as the ion product constant for water and is symbolized by K_w.

$$K_w = [H^+][OH^-] \qquad (7\text{-}11)$$

At 25°C, K_w is equal to 1×10^{-14}. At 25°C the product of the hydrogen ion concentration times the hydroxide ion concentration must always equal 1×10^{-14}.

$$[H^+][OH^-] = 1 \times 10^{-14} \text{ (at 25°C)} \qquad (7\text{-}12)$$

A practical consequence of this is that since the product of $[H^+]$ and $[OH^-]$ is a constant, then the one factor must of necessity decrease when the other increases.

Equation (7-12), however, is true only at 25°C; whereas, Equation (7-11) is true at all possible temperatures. The degree of ionization as shown in Equation (7-9) increases with an increase in temperature. Thus, at 100°C, K_w is approximately 1×10^{-12}.

We can, therefore, state correctly that at any temperature the hydrogen ion concentration times the hydroxide ion concentration will equal the ion product constant for water at that temperature. It would be incorrect, however, to state that under all conditions the product of the hydrogen ion concentration and the hydroxide ion concentration is equal to 1×10^{-14}.

Definition of pH

The pH of a solution was defined by Sørensen, in 1909, as either the negative logarithm of the hydrogen ion activity or the logarithm of one over the hydrogen ion activity. Equations (7-13) and (7-14) represent this mathematically.

$$pH = -\log a_{H^+} \qquad (7\text{-}13)$$

$$pH = \log \frac{1}{a_{H^+}} \qquad (7\text{-}14)$$

As explained in Chapter 1, in dilute solutions the activity of an ionic species approximates the concentration of the species. For simplicity then, we can rewrite Equations (7-13) and (7-14) as

$$pH = -\log [H^+] \qquad (7\text{-}15)$$

$$pH = \log \frac{1}{[H^+]} \qquad (7\text{-}16)$$

Throughout the remainder of this chapter, the term "activity" will be dispensed with and "concentration" will be used.

pH Values

The convenience of the expression "pH value" can be illustrated as follows:

A solution contains 0.00003 gram ion (or mole) of hydrogen ion per liter. This can be expressed as 3×10^{-5} m/l, but it can also be expressed as having a pH value equal to 4.52 as shown in Example 7-1.

EXAMPLE 7-1: Calculate the pH of a $0.00003M$ solution of hydrogen ion.

$$[H^+] = 3 \times 10^{-5}$$

$$pH = -\log (3 \times 10^{-5})$$

$$\log (3 \times 10^{-5}) = \log 3 + \log 10^{-5}$$

$$= 0.4771 - 5.0000$$

$$= -4.5229$$

$$pH = 4.5229 = 4.52$$

The calculation can also be done by expressing the term 3×10^{-5} entirely as a power of 10. The logarithm of 3 is 0.4771 meaning that 3 is equal to 10 raised to the 0.4771 power. Therefore,

$$3 \times 10^{-5} = 10^{.4771} \times 10^{-5} = 10^{-4.5229}$$

$$pH = 4.5229 = 4.52$$

You will also note that the pH has been rounded off to two decimal places. This is the customary manner of reporting pH. Rarely does data justify expression of pH past the second decimal place.

Example 7-2 shows the reverse calculation from pH value to hydrogen ion concentration.

> **EXAMPLE 7-2:** Calculate the hydrogen ion concentration in a solution having a pH of 3.45.

$$pH = 3.45$$

$$[H^+] = 10^{-3.45} = 10^{.55} \times 10^{-4}$$

$$[H^+] = 3.6 \times 10^{-4}$$

Acidic, Basic, or Neutral Solution

By definition, a neutral solution is one in which the hydrogen ion concentration is equal to the hydroxide ion concentration. An acidic solution is one in which the hydrogen ion concentration is greater than the hydroxide ion concentration. A basic solution is one in which the hydrogen ion concentration is less than the hydroxide ion concentration.

$$[H^+] = [OH^-] \quad \text{neutral} \qquad \qquad \textbf{(7-17)}$$

$$[H^+] > [OH^-] \quad \text{acidic} \qquad \qquad \textbf{(7-18)}$$

$$[H^+] < [OH^-] \quad \text{basic} \qquad \qquad \textbf{(7-19)}$$

Therefore, in a neutral solution at 25° C the mathematical solution of Equation (7-12) shows that the hydrogen ion concentration is equal to 1×10^{-7}.

$$[H^+] = [OH^-]$$

$$[H^+]^2 = 1 \times 10^{-14}$$

$$[H^+] = 1 \times 10^{-7}$$

The pH of a neutral solution at 25°C would therefore be 7. Thus, at 25°C we can write the following:

For a neutral solution

$$pH = 7 \qquad \qquad \textbf{(7-20)}$$

and

$$pH = pOH \qquad \textbf{(7-21)}$$

For an acidic solution

$$pH < 7 \qquad \textbf{(7-22)}$$

and

$$pH < pOH \qquad \textbf{(7-23)}$$

For a basic solution

$$pH > 7 \qquad \textbf{(7-24)}$$

and

$$pH > pOH \qquad \textbf{(7-25)}$$

Note that in Equation (7-18) an acidic solution is one in which the $[H^+]$ is *greater* than the $[OH^-]$, but in Equation (7-22), an acidic solution is one in which the pH is *less* than 7 or *less* than pOH. This apparent contradiction occurs because pH is the *negative* logarithm of the hydrogen ion concentration.

Equations (7-17), (7-18), (7-19), (7-21), (7-23), and (7-25) are valid for any aqueous solution and at any temperature. Equations (7-20), (7-22), and (7-24), however, are only true at 25°C. Some textbooks incorrectly state without any conditional requirements that a neutral solution has a pH of 7. The following example will illustrate why this should not be done.

EXAMPLE 7-3: At 50°C, K_w is equal to 6.30×10^{-14}. Is a solution having a pH value of 6.70 at 50°C neutral, basic, or acidic?

$$pH = 6.70$$

$$[H^+] = 10^{-6.70} = 10^{.30} \times 10^{-7}$$

$$= 2.0 \times 10^{-7}$$

$$[H^+][OH^-] = 6.30 \times 10^{-14}$$

$$[OH^-] = \frac{6.30 \times 10^{-14}}{2.0 \times 10^{-7}} = 3.15 \times 10^{-7}$$

Therefore, at pH = 6.70 the solution is basic because $[H^+] <$ $[OH^-]$. Another solution to this problem is to calculate pH and pOH.

$$K_w = 6.30 \times 10^{-14} = 10^{.80} \times 10^{-14}$$

$$pK_w = 13.2$$

$$pH + pOH = pK_w$$

$$pH = 6.70$$

$$pOH = 6.50$$

Once again we would conclude the solution is basic because pH > pOH.

Strengths of Electrolytes in Water

An electrolyte is a substance which ionizes in water. Electrolytes exist principally as one of two types: strong electrolytes—those which exist essentially in the ionized form and for calculation purposes are considered 100% ionized; and weak electrolytes—those which exist essentially in the undissociated form and only ionize to a limited and small but definite extent.

Examples of strong electrolytes include hydrochloric acid, nitric acid, sulfuric acid, sodium hydroxide, and potassium hydroxide among others. Examples of weak electrolytes include phosphoric acid, acetic acid (as well as most organic acids), ammonium hydroxide, and aniline (as well as most organic bases). A strong acid (or base) refers to its degree of ionization and thus is considered a strong acid (or base) regardless of how weak in concentration (moles per liter) it is! Conversely, a weak acid (or base) refers to an electrolyte which ionizes only slightly in solution and is considered a weak acid (or base) regardless of how strong in concentration (moles per liter) it is! In fact, the higher the concentration of a strong acid or base, the less ionized the substance becomes, and the lower the concentration of a weak acid or base, the more ionized the substance becomes. The neutralization of 100 ml of $0.1M$ solution of a weak monobasic acid will require the same amount of base as is required for 100 ml of $0.1M$ solution of a strong monobasic acid.

With the preceding information in mind, we can now direct ourselves to the specific calculations of pH in various types of acidic and basic solutions. *For all calculations we will assume a temperature of 25°C.*

Case I—Solutions of Strong Acids

The simplest calculation of pH is that involving solutions of strong acids. If the degree of ionization is assumed to be 100%, the $[H^+]$ in equivalents (or moles) per liter is numerically equal to the normality of the solution. The following two examples will illustrate:

EXAMPLE 7-4: Calculate the pH of a solution of hydrochloric acid which contains 5.25 g HCl in 250 ml.
First the normality of the solution is determined.

$$N = \frac{5.25\,g \times \dfrac{1000\,ml/l}{250\,ml}}{36.5\,g/equiv} = 0.575N$$

Since HCl is assumed to be 100% ionized, $[H^+] = 0.575$.

$$[H^+] = 5.75 \times 10^{-1} = 10^{.76} \times 10^{-1} = 10^{-.24}$$

$$pH = 0.24$$

EXAMPLE 7-5: Calculate the pH of a $0.020M$ solution of sulfuric acid.
Since H_2SO_4 has 2 replaceable hydrogen ions, the normality is twice the molarity.

$$[H^+] = 0.040 = 4.0 \times 10^{-2}$$
$$= 10^{.60} \times 10^{-2} = 10^{-1.40}$$
$$pH = 1.40$$

Case II—Solutions of Strong Bases

The calculation of pH in solutions of strong bases is also straightforward. Again, if the assumption is made that ionization is 100% complete, the $[OH^-]$ in equivalents per liter is numerically equal to the normality of the solution. The pH value can be calculated by either of two methods. Example 7-6 will show the calculation where the $[H^+]$ is obtained from the $[OH^-]$ by means of K_w. Example 7-7 will show the calculation involving pOH values. Again both examples assume a temperature of 25°C.

EXAMPLE 7-6: Calculate the pH of a solution which contains 2.00 g of NaOH in 2.000 liters of solution.

$$\text{no equiv} = \frac{2.00\,g}{40.0\,g/\text{equiv}} = 0.05\,\text{equiv}$$

$$N = \text{equiv/liter} = \frac{0.05\,\text{equiv}}{2\,\text{liters}} = 0.025\,N$$

$$[OH^-] = 0.025$$

$$[H^+] = \frac{1.0 \times 10^{-14}}{2.5 \times 10^{-2}} = 4.0 \times 10^{-13}$$

$$[H^+] = 10^{.60} \times 10^{-13} = 10^{-12.40}$$

$$pH = 12.40$$

EXAMPLE 7-7: Calculate the pH of a $0.0050\,N$ solution of $Ca(OH)_2$.

$$[OH^-] = 5.0 \times 10^{-3} = 10^{.70} \times 10^{-3} = 10^{-2.30}$$

$$pOH = 2.30$$

$$pH + pOH = pK_w = 14$$

$$pH = 14 - 2.30$$

$$pH = 11.70$$

Fundamentally the pH calculation in basic solutions as per Examples 7-6 and 7-7 are equally valid. The mathematics involved in Example 7-7 are easier and therefore less likely to involve a computational error. The basic difference is that in Example 7-6, a division was required, whereas in Example 7-7, a subtraction was required. Both examples required the computation of a logarithm.

It will be shown in Chapter 8, Example 8-2, that this straightforward approach is not applicable in very dilute solutions of strong acids and strong bases.

Case III—Solutions of Weak Acids (Monobasic)

A weak acid is an example of a weak electrolyte. Since ionization occurs only to a small extent, the hydrogen ion concentration is not equal to the molarity of the acid. This complicates the calculation of pH values. A weak acid which we can designate as HA will dissociate according to Equation (7-26) or (7-27).

$$HA \rightleftharpoons H^+ + A^-$$ (7-26)

or

$$HA + H_2O \rightleftharpoons H_3O^+ + A^-$$ (7-27)

where A^- represents the anion of the weak acid. Equation (7-27) represents the dissociation as a Brønsted-Lowry reaction and the reason that dissociation occurs only to a small extent is that H_3O^+ (acid$_2$) and A^- (base$_1$) are stronger species than HA (acid$_1$) and H_2O (base$_2$). The equilibrium represented in Equation (7-26) can be expressed by applying the Law of Mass Action as in Equation (7-28).

$$K_a = \frac{[H^+][A^-]}{[HA]}$$ (7-28)

The constant K_a is a specialized constant known as the dissociation (or ionization) constant of the acid. The values of K_a for several weak acids are listed in Appendix B.

If the solution contains only the weak acid and no other source of hydrogen ions or anion of the acid, Equation (7-26) shows that equal molar concentrations of $[H^+]$ and $[A^-]$ will be present at equilibrium since one of each is formed by the dissociation reaction. Actually, there is an additional source of hydrogen ions in an aqueous solution. The solvent water furnishes a small amount of $[H^+]$ due to self-ionization. However, in an acid solution, the dissociation of water is strongly repressed and unless the acid is extremely weak in strength or extremely low in concentration, the contribution of $[H^+]$ from water can be safely ignored. Thus,

$$[H^+] = [A^-]$$ (7-29)

If the total analytical concentration of the weak acid (both associated and ionized forms) is known, it can be designated as C_a, where C_a is in moles per liter.

At equilibrium, the concentration of undissociated HA would be equal to $C_a - [H^+]$ since $[H^+]$ is equal to the number of moles per liter of the acid which dissociated.

$$[HA]_{eq} = C_a - [H^+] \qquad (7\text{-}30)$$

Combining Equations (7-29) and (7-30) into Equation (7-28), we have

$$K_a = \frac{[H^+]^2}{C_a - [H^+]} \qquad (7\text{-}31)$$

Equation (7-31) can be simplified if we make the assumption that $[H^+]$ is small and negligible in comparison with C_a. This will generally be true for weak acids at a concentration of $0.01M$ or greater. The assumption should always be made, but if $[H^+]$ calculated using the assumption amounts to approximately 5% or more of C_a, then Equation (7-31) must be solved using the quadratic equation.

$$C_a - [H^+] \cong C_a \quad \text{if} \quad [H^+] \lll C_a$$

Thus,

$$K_a = \frac{[H^+]^2}{C_a}$$

or

$$[H^+]^2 = K_a C_a$$

and

$$[H^+] = \sqrt{K_a C_a} \qquad (7\text{-}32)$$

Equation (7-32) is the general formula used for calculating the pH value of solutions of weak acids. Example 7-8 will illustrate the use of Equation (7-32).

EXAMPLE 7-8: Calculate the pH of a $0.080M$ solution of acetic acid. K_a for acetic acid is 1.8×10^{-5}.

$$HOAc \rightleftharpoons H^+ + OAc^-$$

$$K_a = \frac{[H^+][OAc^-]}{[HOAc]} = 1.8 \times 10^{-5}$$

$$1.8 \times 10^{-5} = \frac{[H^+]^2}{0.080 - [H^+]}$$

but

$$[H^+] \lll 0.080$$

therefore,

$$[H^+]^2 = (1.8 \times 10^{-5})(0.080)$$
$$[H^+] = \sqrt{1.44 \times 10^{-6}} = 1.2 \times 10^{-3}$$

We see that this value could have been calculated directly from Equation (7-32).

$$[H^+] = \sqrt{K_a C_a}$$
$$= \sqrt{(1.8 \times 10^{-5})(0.08)}$$
$$= 1.2 \times 10^{-3}$$
$$= 10^{.08} \times 10^{-3} = 10^{-2.92}$$
$$pH = 2.92$$

Case IV—Solutions of Weak Bases (Monoacidic)

The calculations of pH values in solutions of weak bases is entirely analogous to that just covered in Case III. Ammonium hydroxide can be used as an example of a weak base. It dissociates as in Equation (7-33) or (7-34).

$$NH_4OH \rightleftharpoons NH_4^+ + OH^- \qquad \text{(7-33)}$$

or

$$NH_3 + H_2O \rightleftharpoons NH_4^+ + OH^- \qquad \text{(7-34)}$$

Again this reaction proceeds only slightly from left to right because NH_4^+ (acid$_2$) is a stronger acid than water (acid$_1$) and OH^- (base$_1$) is a stronger base than NH_3 (base$_2$).

The equilibrium involved in Equation (7-33) can be expressed as in Equation (7-35).

$$K_b = \frac{[NH_4^+][OH^-]}{[NH_4OH]} \qquad \text{(7-35)}$$

where K_b indicates the dissociation (or ionization) constant of the base.

Again if the solution contains only the weak base and no other source of hydroxide ion other than the self-ionization of water, and contains no other source of the cation of the base, Equation (7-33) will show us that

$$[NH_4^+] = [OH^-]$$

Again the total analytical concentration of the base can be designated as C_b. If this is done, the equilibrium concentration of undissociated NH_4OH is $C_b - [OH^-]$.

Therefore,

$$K_b = \frac{[OH^-]^2}{C_b - [OH^-]}$$

which will simplify if $[OH^-] \lll C_b$ to

$$K_b = \frac{[OH^-]^2}{C_b}$$

$$[OH^-] = \sqrt{K_b C_b} \tag{7-36}$$

It should now be noted that Equation (7-36) is analogous to Equation (7-32).

Examples 7-9 and 7-10 illustrate pH calculations involved in a solution of a weak base.

EXAMPLE 7-9: Calculate the pH value of a $0.010M$ solution of ammonium hydroxide. K_b for ammonium hydroxide equals 1.8×10^{-5}.

$$[OH^-] = \sqrt{K_b C_b}$$
$$= \sqrt{(1.8 \times 10^{-5})(1.0 \times 10^{-2})} = \sqrt{18 \times 10^{-8}}$$
$$= 4.25 \times 10^{-4} = 10^{.63} \times 10^{-4} = 10^{-3.37}$$

$$pOH = 3.37$$
$$pH = 14.00 - 3.37$$
$$= 10.63$$

EXAMPLE 7-10: Calculate the pH value of a $0.10M$ solution of ethylamine, $C_2H_5NH_2$. $K_b = 5.4 \times 10^{-4}$. To illustrate the equilibrium involved, Equation (7-37) is shown.

$$C_2H_5NH_2 + H_2O \rightleftharpoons C_2H_5NH_3^+ + OH^- \qquad (7\text{-}37)$$

and the equilibrium constant expression is shown in Equation (7-38).

$$K_b = \frac{[C_2H_5NH_3^+][OH^-]}{[C_2H_5NH_2]} = 5.4 \times 10^{-4} \qquad (7\text{-}38)$$

Application of Equation (7-36) gives

$$[OH^-] = \sqrt{(5.4 \times 10^{-4})(1 \times 10^{-1})}$$

$$= 7.35 \times 10^{-3}$$

$$[H^+] = \frac{1 \times 10^{-14}}{7.35 \times 10^{-3}} = 1.36 \times 10^{-12}$$

$$= 10^{.13} \times 10^{-12} = 10^{-11.87}$$

$$pH = 11.87$$

The preceding two examples again illustrate the two methods for calculating pH values from $[OH^-]$. Example 7-9 calculates pOH and then subtraction from 14 gives pH. Example 7-10 calculates the $[H^+]$ from $[OH^-]$ and K_w and then computes the pH.

Buffer Solutions

The preceding two topics (Case III and Case IV) illustrate the equilibria involved in pure solutions of weak acids and weak bases. The next more complicated situation is in a solution containing not only the weak electrolyte but also a salt of this same electrolyte. A solution such as this is known as a buffer solution. The anion of the salt of the weak electrolyte is also the conjugate base of the acid, and so a buffer can also be defined as a solution containing a conjugate acid-base pair. The word buffer is an adaptation of the German word "puffer," meaning pad or cushion. A buffer solution is, therefore, a solution which acts as a pad or cushion against the effect of addition of both acids and bases. A buffer solution is not a solution in which the pH does not change upon the addition of acid or base but is a solution in which the pH does not change as much as it would if the same amount of acid or base were added to water. For example, if a drop of $0.2N$ HCl is added to a liter of pure water at pH 7, the resulting solution will change to approximately pH 5. Likewise, a similar amount of NaOH would change it to about pH 9. However, if the same quantity of acid or base were added to a neutral solution containing a mixture of phosphate salts, the change in pH would be almost negligible.

A simple effective buffer solution can be prepared by mixing equal molar amounts of acetic acid with sodium acetate. The resulting solution will react with both acids and bases as will be shown in the following discussion. Acetic acid ionizes only to a limited extent as was seen earlier in Equation (7-2).

$$HOAc \rightleftharpoons H^+ + OAc^- \qquad (7\text{-}2)$$

On the other hand, sodium acetate is a strong electrolyte and exists essentially 100% in the ionized form.

$$NaOAc \rightarrow Na^+ + OAc^- \qquad (7\text{-}39)$$

Consideration of Equations (7-2) and (7-39) along with general knowledge of equilibria will show in a qualitative way why a buffer solution resists changes in pH.

First, consider what happens when a strong acid is added. The hydrogen ions from the strong acid will react with the acetate ions from the salt [Equation (7-39)],

$$NaOAc \rightarrow Na^+ + \boxed{\begin{array}{c} OAc^- \\ + \\ H^+ \end{array}} \rightleftharpoons HOAc$$

forming additional undissociated free acetic acid. The equilibrium in Equation (7-2) lies far to the left, thereby, the bulk of the H^+ are effectively removed from the solution. It must be noted that not all the added H^+ from the strong acid are removed and, therefore, the pH will drop slightly. The section on buffer capacity will show quantitatively what happens when an acid is added to a buffer solution.

What happens in a buffer solution when a strong base is added? The hydroxide ions react with the H^+ from Equation (7-2),

$$HOAc \rightleftharpoons \boxed{\begin{array}{c} H^+ \\ + \\ OH^- \end{array}} + OAc^- \\ \rightleftharpoons H_2O$$

forming undissociated water. This disturbs the equilibrium expressed in Equation (7-2) and additional acetic acid will dissociate to restore the equilibrium, thereby, again furnishing a supply of H^+. The extent to which this can continue depends on the original concentration of the weak acid.

In summary, the buffering action is due to the fact that the equilibrium in Equation (7-2) shifts to the left when acid is added and to the right when base is added. Added hydrogen ions react with the excess acetate ions furnished by the salt, while added hydroxide ions are neutralized by hydrogen ions released by the further dissociation of acetic acid. The net result is that as long as there is an excess of acetate ions or undissociated acetic acid over the amount of acid or base added, the pH will change only slightly.

Case V—Solutions of Weak Acids and Salts of the Weak Acid

The equation for the calculation of pH values of a solution of a weak acid and its salt is a direct application of the law of mass action.

$$K_a = \frac{[H^+][A^-]}{[HA]} \qquad (7\text{-}28)$$

Let: C_a = original analytical concentration of weak acid in moles/liter
C_s = original concentration of salt in moles/liter

$$HA \rightleftharpoons H^+ + A^- \qquad (7\text{-}26)$$

$$NaA \rightarrow Na^+ + A^- \qquad (7\text{-}40)$$

At equilibrium, [HA] will be equal to the original concentration C_a less that which ionizes. The concentration of HA which ionizes is seen from Equation (7-26) to be equal to $[H^+]$ since each mole of HA produces one mole of H^+.

$$[HA]_{eq} = C_a - [H^+] \qquad (7\text{-}41)$$

Also at equilibrium, $[A^-]$ will be equal to the original concentration C_s plus that which is formed by the dissociation of the acid. Again the $[A^-]$ formed by dissociation of the acid is equal to $[H^+]$ since they are formed at a 1 : 1 mole ratio as shown in Equation (7-26).

$$[A^-]_{eq} = C_s + [H^+] \qquad (7\text{-}42)$$

Substituting Equations (7-41) and (7-42) into Equation (7-28) gives

$$K_a = \frac{[H^+](C_s + [H^+])}{C_a - [H^+]}$$

rearranging

$$[H^+] = \frac{K_a(C_a - [H^+])}{C_s + [H^+]} \qquad (7\text{-}43)$$

This somewhat cumbersome quadratic equation can be reduced by making the valid assumption that $[H^+]$ is very small compared to both C_a and C_s.

$$[H^+] \lll C_a \quad \text{or} \quad C_s$$

Equation (7-43) then simplifies to

$$[H^+] = \frac{K_a C_a}{C_s} \qquad (7\text{-}44)$$

The relatively simple Equation (7-44) can be used to calculate the pH value of any buffer of a weak acid and its salt. Since the hydrogen ion concentration depends upon the *ratio* of acid concentration to salt concentration, dilution of a buffer has a negligible effect on pH.

A buffer can be prepared by mixing a weak acid and its salt. An example of the calculation involved is shown in Example 7-11. A buffer can also be prepared by the partial neutralization of a weak acid with a strong base. Example 7-12 will illustrate this calculation. An additional method of preparation of a buffer is by the addition of a strong acid to the salt of a weak acid. Example 7-13 will illustrate this calculation.

EXAMPLE 7-11: Calculate the pH of a solution prepared by mixing 100 ml of 0.050M molar acetic acid with 0.82 g of sodium acetate. K_a for acetic acid is 1.8×10^{-5}. Assume no volume change with dissolving of the salt.

$$[NaOAc] = \frac{0.82\,g}{82.03\,g/moles} \times 10 = 0.10M$$

$$[HOAc] = 0.050M$$

$$[H^+] = \frac{(1.8 \times 10^{-5})(0.050)}{0.10}$$

$$= 9 \times 10^{-6} = 10^{.95} \times 10^{-6} = 10^{-5.05}$$

$$pH = 5.05$$

EXAMPLE 7-12: Calculate the pH of a solution prepared by mixing 75 ml of 0.20M acetic acid with 25 ml of 0.40M sodium hydroxide.

$$75\,ml \times 0.20M = 15\,millimoles\,HOAc$$

$$25\,ml \times 0.40M = 10\,millimoles\,NaOH$$

Obviously, the 10 millimoles NaOH will neutralize 10 millimoles of the HOAc forming in the process 10 millimoles of sodium acetate and leaving an excess of 5 millimoles of acetic acid.

$$[HOAc] = \frac{5\,mmoles}{100\,ml} = 0.05M$$

$$[OAc^-] = \frac{10\,mmoles}{100\,ml} = 0.10M$$

$$[H^+] = \frac{(1.8 \times 10^{-5})(0.05)}{0.10} = 9 \times 10^{-6}$$

$$= 10^{.95} \times 10^{-6} = 10^{-5.05}$$

$$pH = 5.05$$

Note that the pH for the last two examples are equal. As a matter of fact, the solutions are identical. It would be impossible to distinguish between them. This illustrates the fact that a buffer can be prepared either by mixing a weak acid with its salt or by the partial neutralization of the acid with a strong base.

EXAMPLE 7-13: Calculate the pH of a solution prepared by mixing 100 ml of 0.050M hydrochloric acid with 1.230 g of sodium acetate. Assume no volume change with the dissolving of the salt.
Originally there are

$$\frac{1.230\,g}{0.0820\,g/mmole} = 15\,millimoles\,NaOAc$$

and

$$100 \, \text{ml} \times 0.05M = 5 \text{ millimoles HCl}$$

The strong acid (HCl) will react with the salt (NaOAc) forming an equivalent amount of undissociated weak acid (HOAc) and salt (NaCl) as shown in Equation (7-45).

$$\text{HCl} + \text{NaOAc} \rightarrow \text{HOAc} + \text{NaCl} \qquad \textbf{(7-45)}$$

Thus, at equilibrium, the solution will contain 5 millimoles HOAc formed by Equation (7-45), 5 millimoles NaCl formed by Equation (7-45), and 10 millimoles NaOAc unreacted,

$$[\text{HOAc}] = \frac{5 \text{ mmoles}}{100 \, \text{ml}} = 0.05M$$

$$[\text{NaOAc}] = \frac{10 \text{ mmoles}}{100 \, \text{ml}} = 0.10M$$

The presence of NaCl does not seriously affect the pH of this solution.

$$[\text{H}^+] = \frac{(1.8 \times 10^{-5})(0.05)}{0.10} = 9 \times 10^{-6}$$

$$[\text{H}^+] = 10^{.95} \times 10^{-6} = 10^{-5.05}$$

$$\text{pH} = 5.05$$

Again note that the pH in Example 7-13 is identical to the pH in Examples 7-11 and 7-12. The only difference in this last solution is the presence of a small amount of sodium chloride. The presence of the sodium chloride does slightly affect the activity coefficients but this effect is small and for our purposes can be safely ignored.

Case VI—Solutions of Weak Bases and Their Salts

The calculations involving weak bases and their salts are identical to those in Case V. The derivation of the equation used is identical and gives

$$[\text{OH}^-] = \frac{K_b C_b}{C_s} \qquad \textbf{(7-46)}$$

EXAMPLE 7-14: Calculate the pH of a solution prepared by adding 50 ml of $0.10M$ HCl to 100 ml of $0.10M$ ethanolamine. K_b for ethanolamine is 2.77×10^{-5}.

At equilibrium, 50 ml of the $0.10M$ ethanolamine would be neutralized.

$$C_b = \frac{50 \times 0.10}{150}$$

$$C_s = \frac{50 \times 0.10}{150}$$

$$[OH^-] = \frac{(2.77 \times 10^{-5}) \dfrac{50 \times 0.10}{150}}{\dfrac{50 \times 0.10}{150}}$$

$$[OH^-] = 2.77 \times 10^{-5} = 10^{.44} \times 10^{-5} = 10^{-4.56}$$

$$pOH = 4.56$$

$$pH = 14.00 - 4.56 = 9.44$$

Two general facts can be gathered from Example 7-14. The first is that it does not matter whether the values for C_b and C_s are in moles per liter or just in moles. The reason for this is that the $[OH^-]$ or $[H^+]$ is proportional to a ratio of the two concentrations and hence, also proportional to the number of moles of the two species since both are in the same solution.

The second fact illustrated by Example 7-14 is that at the midpoint of a titration of a weak base (or acid) the $[OH^-]$ (or $[H^+]$) is equal to the ionization constant.

Buffer Capacity

The amount of acid or base that a buffered solution can handle with only a small change in pH is called its *buffer capacity*. Buffer capacity is defined in a quantitative manner as the number of equivalents of a strong base needed to raise the pH of one liter of solution by one pH unit. If a high buffer capacity is desired, this can be achieved by using high concentrations of acid and salt. For a given total concentration of acid and salt, the greatest buffer capacity occurs when the concentration of the acid and salt are equimolar. This situation occurs at the midpoint in the titration of a weak acid.

The following examples illustrate the type of calculations involved:

EXAMPLE 7-15: Calculate the ratio in which acetic acid and sodium acetate must be mixed to give a buffer having a pH 4.50. The [H^+] is calculated from the pH.

$$[H^+] = 10^{-4.50} = 10^{.50} \times 10^{-5}$$
$$= 3.2 \times 10^{-5}$$

This [H^+] is placed into the K_a expression.

$$K_a = \frac{[H^+][OAc^-]}{[HOAc]}$$

$$1.8 \times 10^{-5} = \frac{(3.2 \times 10^{-5})[OAc^-]}{[HOAc]}$$

$$\frac{[OAc^-]}{[HOAc]} = \frac{1.8 \times 10^{-5}}{3.2 \times 10^{-5}} = 0.56$$

Thus, sodium acetate and acetic acid should be mixed in a molar ratio of 0.56 to 1. The dilution volume is not critical.

EXAMPLE 7-16: How many ml of $0.100M$ NaOH must be added to 50.0 ml of $0.100M$ acetic acid to give a buffer solution having a pH 4.50?

From Example 7-15, it is seen that there must be 0.56 moles of acetate ion for each mole of acetic acid. The addition of sodium hydroxide to acetic acid converts part of the acid to sodium acetate.

$$HOAc + NaOH \rightarrow H_2O + NaOAc$$

Because the sodium acetate is formed by neutralization of the acetic acid, the total of the acetate ion formed and the acetic acid remaining must equal 5.00 millimoles.

Let X = millimoles of acetate ion formed, then $5.00 - X$ = millimoles of acetic acid remaining.

$$\frac{[OAc^-]}{[HOAc]} = 0.56$$

Therefore,

$$\frac{X}{5.00 - X} = 0.56$$

$$X = 1.79$$

In order to form 1.79 millimoles of acetate ion, 1.79 millimoles of sodium hydroxide must be added. Therefore, 17.9 ml of 0.100 M NaOH must be added.

EXAMPLE 7-17: Calculate the buffer capacity of a solution containing 0.125 M NaOAc and 0.125 M HOAc.

The [H$^+$] in the solution is equal to K_a.

$$[H^+] = \frac{1.8 \times 10^{-5} \times 0.125}{0.125} = 1.8 \times 10^{-5}$$

To calculate the buffer capacity, the pH of the solution must be raised 1 pH unit.

original [H$^+$] = 1.8×10^{-5} = $10^{.25} \times 10^{-5}$ = $10^{-4.75}$

original pH = 4.75

desired pH = 5.75

desired [H$^+$] = $10^{-5.75}$ = $10^{.25} \times 10^{-6}$ = 1.8×10^{-6}

Note that the raising of the pH value by one unit lowers the [H$^+$] by exactly one power of ten. Therefore, it is unnecessary to convert from [H$^+$] to pH and then from pH + 1 back to [H$^+$].

Buffer capacity is defined per liter of solution, so assume that one liter of buffer solution is available. One liter of buffer contains 0.125 moles of acetate ion and 0.125 moles of acetic acid. The addition of sodium hydroxide converts part of the acetic acid into acetate ion.

$$[H^+] = \frac{K_a C_a}{C_s}$$

Let X = moles of sodium hydroxide added, therefore,

$$[H^+] = 1.8 \times 10^{-6}$$

$$K_a = 1.8 \times 10^{-5}$$

$$C_a = 0.125 - X \quad \text{(assuming no change in volume)}$$

$$C_s = 0.125 + X$$

$$1.8 \times 10^{-6} = \frac{1.8 \times 10^{-5}(0.125 - X)}{0.125 + X}$$

$$\frac{1.8 \times 10^{-6}}{1.8 \times 10^{-5}} = \frac{0.125 - X}{0.125 + X}$$

$$0.10 = \frac{0.125 - X}{0.125 + X}$$

$$0.0125 + 0.1X = 0.125 - X$$

$$1.10X = 0.1125$$

$$X = 0.102$$

The buffer capacity of the solution is therefore 0.102 moles per liter.

The buffer capacity for some other buffer concentrations are shown in Table 7-1.

Table 7-1

Buffer Capacities of Some Sodium Acetate—Acetic Acid Buffers

Concentration of HOAc, M	Concentration of NaOAc, M	Initial pH	Buffer capacity m/l
1.0	1.0	4.75	0.816
0.5	0.5	4.75	0.408
0.125	0.125	4.75	0.102
0.05	0.05	4.75	0.041
0.10	0.05	4.44	0.075
0.05	0.10	5.05	0.043
0.10	0.50	5.44	0.088
0.50	0.10	4.05	0.300

When the salt concentration is greater than the acid concentration, the buffer is more effective in resisting a pH change with the addition of acid than with the addition of base. Conversely, if the acid concentration is greater than the salt concentration, the buffer is more effective in resisting the addition of base.

Salts of Weak Acids and Weak Bases

A salt formed from the neutralization of a weak acid with a strong base or from the neutralization of a weak base with a strong acid is a strong electrolyte. That is, it dissociates essentially 100%. The ion of the weak acid or base, however, reacts with water to form some of the undissociated weak electrolyte. In forming the undissociated weak electrolyte, an equivalent amount of free hydrogen or hydroxide ions are formed. An example of this reaction using sodium acetate is shown in Equations (7-39) and (7-47).

$$NaOAc \rightarrow Na^+ + OAc^- \qquad (7\text{-}39)$$

$$\cancel{Na^+} + OAc^- + H_2O \rightleftharpoons \cancel{Na^+} + HOAc + OH^- \qquad (7\text{-}47)$$

Therefore, the resulting solution is not neutral. A salt of a weak acid and a strong base such as sodium acetate yields a basic solution when dissolved in water. A salt of a weak base and a strong acid such as ammonium chloride yields an acidic solution, when dissolved in water [Equation (7-48)]

$$NH_4^+ + \cancel{Cl^-} + H_2O \rightleftharpoons \cancel{Cl^-} + NH_4OH + H^+ \qquad (7\text{-}48)$$

The term for the type of reaction illustrated by Equations (7-47) and (7-48) is known as *hydrolysis* and is a type of Brønsted-Lowry acid-base reaction. This type of reaction does not occur when the salt of a strong acid and a strong base such as NaCl is dissolved in water.

The derivation of the general equation used in calculating pH values for solution of salts which undergo hydrolysis will be illustrated in Case VII and Case VIII.

Case VII—Solution of a Salt of a Weak Acid and a Strong Base

The derivation of the general formula used for calculating pH values in solutions of salts of weak acids and strong bases will be illustrated for sodium acetate.

The reaction is as follows:

$$\cancel{Na^+} + OAc^- + H_2O \rightleftharpoons \cancel{Na^+} + HOAc + OH^- \qquad (7\text{-}47)$$

Writing the equilibrium constant,

$$K'_{hyd} = \frac{[Na^+][HOAc][OH^-]}{[Na^+][OAc^-][H_2O]} \qquad (7\text{-}49)$$

The concentration of sodium ions is identical on both sides of the equation and, hence, will cancel out. The concentration of undissociated water is so large that it remains essentially constant and can be included in the hydrolysis constant.
Therefore,

$$K_{hyd} = \frac{[HOAc][OH^-]}{[OAc^-]} \qquad (7\text{-}50)$$

Multiplying both the numerator and denominator of Equation (7-50) by $[H^+]$ yields

$$K_{hyd} = \frac{[HOAc][OH^-][H^+]}{[OAc^-][H^+]} \qquad (7\text{-}51)$$

It will be noted that the last two terms of the numerator in Equation (7-51) equals K_w.

$$[OH^-][H^+] = K_w$$

The other three terms are equal to $1/K_a$.

$$\frac{[HOAc]}{[OAc^-][H^+]} = \frac{1}{K_a}$$

Therefore,

$$K_{hyd} = \frac{K_w}{K_a} \qquad (7\text{-}52)$$

The hydrolysis constant therefore depends upon the ratio of the two constants K_w and K_a.
Combining Equations (7-52) and (7-50) yields

$$K_{hyd} = \frac{K_w}{K_a} = \frac{[HOAc][OH^-]}{[OAc^-]} \qquad (7\text{-}53)$$

We can see from Equation (7-47) that $[OH^-] = [HOAc]$ since they form on a one mole to one mole basis. Also, if C_s equals the original concentration of the salt in moles/liter, at equilibrium the concentration of acetate ion would be $C_s - [OH^-]$.

Thus,

$$\frac{K_w}{K_a} = \frac{[OH^-]^2}{C_s - [OH^-]}$$

Since $[OH^-]$ is very small compared to C_s this simplifies to

$$\frac{K_w}{K_a} = \frac{[OH^-]^2}{C_s}$$

or

$$[OH^-] = \sqrt{\frac{K_w C_s}{K_a}} \qquad (7\text{-}54)$$

To solve for $[H^+]$, substitute $K_w/[H^+]$ for $[OH^-]$.

$$\frac{K_w}{[H^+]} = \sqrt{\frac{K_w C_s}{K_a}}$$

$$[H^+] = \sqrt{\frac{K_w^2}{\frac{K_w C_s}{K_a}}}$$

$$[H^+] = \sqrt{\frac{K_w K_a}{C_s}} \qquad (7\text{-}55)$$

Equation (7-55) is an easier equation to remember than Equation (7-54) as will be illustrated in the Summary.

An alternate method for deriving an equation to solve the hydrogen ion concentration in a solution of a salt of a weak acid is based upon the Brønsted-Lowry theory. The anion of an acid is the conjugate base of that acid.

$$HOAc + H_2O \rightleftharpoons H_3O^+ + OAc^- \qquad (7\text{-}6)$$

The conjugate base (OAc^-) can react slightly with water as was shown in Equation (7-47).

$$OAc^- + H_2O \rightleftharpoons HOAc + OH^- \qquad (7\text{-}47)$$

K_a for acetic acid is obtained from Equation (7-6).

$$K_a = \frac{[H^+][OAc^-]}{[HOAc]} \qquad (7\text{-}56)$$

K_b for the acetate ion is obtained from Equation (7-47).

$$K_b = \frac{[HOAc][OH^-]}{[OAc^-]} \qquad (7\text{-}57)$$

The product of $K_a K_b$ for any conjugate acid-base pair is equal to K_w as seen from Equations (7-56) and (7-57).

$$K_a K_b = \frac{[H^+][OAc^-]}{[HOAc]} \times \frac{[HOAc][OH^-]}{[OAc^-]} = [H^+][OH^-]$$
$$= K_w$$

Therefore, $K_b = (K_w/K_a)$, and it follows that the hydrolysis constant given in Equation (7-52) is actually K_b. The hydroxide ion concentration in the solution can now be calculated as it would for a solution of any weak base [Equation (7-36)].

EXAMPLE 7-18: Calculate the pH of a solution prepared by dissolving 0.82 g sodium acetate in 100 ml of water.

$$[OAc^-] = \frac{0.82\ g/100\ ml}{82.03\ g/mole} \times 10 = 0.10\ mole/liter$$

Using Equation (7-55)

$$[H^+] = \sqrt{\frac{K_w K_a}{C_s}} = \sqrt{\frac{(1 \times 10^{-14})(1.8 \times 10^{-5})}{0.10}}$$
$$= \sqrt{1.8 \times 10^{-18}}$$
$$[H^+] = 1.35 \times 10^{-9} = 10^{.13} \times 10^{-9} = 10^{-8.87}$$
$$pH = 8.87$$

Using Equation (7-36)

$$[OH^-] = \sqrt{K_b C_b}$$

$$K_b = \frac{K_w}{K_a} = \frac{1 \times 10^{-14}}{1.8 \times 10^{-5}} = 5.6 \times 10^{-10}$$

$$[OH^-] = \sqrt{(5.6 \times 10^{-10})(0.10)} = 7.5 \times 10^{-6}$$

$$= 10^{.87} \times 10^{-6} = 10^{-5.13}$$

$$pOH = 5.13$$

$$pH = 14 - 5.13 = 8.87$$

EXAMPLE 7-19: Calculate the pH of a solution prepared by titrating 50.0 ml of $0.20M$ acetic acid with 50.0 ml of $0.2N$ sodium hydroxide.

$$HOAc + NaOH \rightarrow NaOAc + H_2O$$

$$50.0 \text{ ml} \times 0.20M = 10.0 \text{ mmoles HOAc}$$

$$50.0 \text{ ml} \times 0.20N = 10.0 \text{ meq NaOH}$$

The acetic acid will be completely neutralized with the formation of 10.0 mmoles NaOAc in 100 ml of solution.

$$C_s = [NaOAc] = \frac{10.0 \text{ mmoles}}{100 \text{ ml}} = 0.10M$$

Using Equation (7-55)

$$[H^+] = \sqrt{\frac{K_w K_a}{C_s}} = \sqrt{\frac{(1 \times 10^{-14})(1.8 \times 10^{-5})}{0.10}}$$

$$[H^+] = 1.35 \times 10^{-9}$$

$$pH = 8.87$$

Using Equation (7-36)

$$[OH^-] = \sqrt{K_b C_b} = \sqrt{(5.6 \times 10^{-10})(0.10)}$$

$$= 7.5 \times 10^{-6}$$

$$pOH = 5.13$$

$$pH = 8.87$$

Again, we see that the pH in Example 7-18 is identical to the pH in Example 7-19. The solutions are identical as it makes no difference whether it was prepared by dissolution of the salt or by formation of the salt by neutralization. Hydrolysis will occur in either case. The practical consequence of hydrolysis is that the end point in the titration of a weak acid or a weak base will not occur at a pH 7. To assume that the solutions were neutral at the equivalence point would cause an appreciable error.

EXAMPLE 7-20:   Calculate the percentage error in the titration of acetic acid if the titration is stopped at pH 7.00.

The equivalence point in the titration of acetic acid occurs at pH 8.87 (see Example 7-19), therefore, at any pH less than this value the solution contains a mixture of the unreacted acetic acid and sodium acetate formed by the neutralization reaction. This results in a buffer solution. The pH of buffer solutions are independent of concentration, therefore, any convenient amount of sample may be assumed for the solving of this example.

Assume 100 millimoles of acetic acid in the sample. At pH 7.00, X millimoles of acetic acid will remain and $100 - X$ millimoles of acetate ion will be formed.

$$[H^+] = \frac{K_a C_a}{C_s}$$

$$1.00 \times 10^{-7} = \frac{1.8 \times 10^{-5}(X)}{100 - X}$$

$$\frac{1.00 \times 10^{-7}}{1.8 \times 10^{-5}} = \frac{X}{100 - X}$$

$$0.0056 = \frac{X}{100 - X}$$

$$X = 0.56$$

Since there were 100 millimoles of acetic acid in the sample and 0.56 millimoles remain at pH 7.00, the error is 0.56%.

Case VIII—Solution of a Salt of a Weak Base and a Strong Acid

The derivation of the equation used for calculating pH values of solutions of salts of weak bases and strong acids is completely analogous to that given in Case VII. The derivation will be repeated for a review. As an example, ammonium chloride will be used.

Consider Equation (7-48)

$$NH_4^+ + \cancel{Cl^-} + H_2O \rightleftharpoons \cancel{Cl^-} + NH_4OH + H^+ \qquad (7\text{-}48)$$

$$K_{hyd} = \frac{[NH_4OH][H^+]}{[NH_4^+]} \qquad (7\text{-}58)$$

Multiply by $[OH^-]/[OH^-]$

$$K_{hyd} = \frac{[NH_4OH][H^+][OH^-]}{[NH_4^+][OH^-]} = \frac{K_w}{K_b} \qquad (7\text{-}59)$$

Combining Equation (7-58) with Equation (7-59) yields

$$\frac{K_w}{K_b} = \frac{[NH_4OH][H^+]}{[NH_4^{+}]}$$

Also, from Equation (7-48)

$$[NH_4OH] = [H^+]$$

and

$$[NH_4^+] = C_s - [H^+]$$

which simplifies to $[NH_4^+] = C_s$ since $[H^+]$ is very small compared to C_s. Therefore,

$$\frac{K_w}{K_b} = \frac{[H^+]^2}{C_s}$$

or

$$[H^+] = \sqrt{\frac{K_w C_s}{K_b}} \qquad (7\text{-}60)$$

Equation (7-60) is analogous to Equation (7-54) and can be used directly to calculate pH. However, to aid in remembering the correct form of the equation, it is advisable to carry the derivation one step further as was done in Case VII.

$$\frac{K_w}{[OH^-]} = \sqrt{\frac{K_w C_s}{K_b}}$$

$$[OH^-] = \sqrt{\frac{K_w K_b}{C_s}} \qquad (7\text{-}61)$$

Equation (7-61) is analogous to Equation (7-55).

Again the Brønsted-Lowry theory can be used to solve for the pH in a solution of the salt of a weak base. It can easily be shown that the K_a for the conjugate acid of ammonium hydroxide (the ammonium ion) is equal to the hydrolysis constant as defined in Equation (7-59). Therefore, the hydrogen ion concentration can be calculated as it would for a solution of any weak acid [Equation (7-32)].

EXAMPLE 7-21:   Calculate the pH of a 0.025M solution of ammonium chloride.

$$K_b = 1.8 \times 10^{-5}$$

Using Equation (7-61)

$$[OH^-] = \sqrt{\frac{(1 \times 10^{-14})(1.8 \times 10^{-5})}{2.5 \times 10^{-2}}}$$

$$= \sqrt{7.2 \times 10^{-18}} = 2.7 \times 10^{-9}$$

$$[H^+] = \frac{1 \times 10^{-14}}{2.7 \times 10^{-9}} = 3.7 \times 10^{-6} = 10^{.57} \times 10^{-6} = 10^{-5.43}$$

$$pH = 5.43$$

Using Equation (7-32)

$$[H^+] = \sqrt{K_a C_a}$$

$$K_a = \frac{K_w}{K_b} = \frac{1 \times 10^{-14}}{1.8 \times 10^{-5}} = 5.6 \times 10^{-10}$$

$$[H^+] = \sqrt{(5.6 \times 10^{-10})(2.5 \times 10^{-2})}$$

$$= 3.7 \times 10^{-6} = 10^{.57} \times 10^{-6} = 10^{-5.43}$$

$$pH = 5.43$$

EXAMPLE 7-22:   Calculate the pH at the endpoint of a titration of 0.100M hydroxylamine with 0.100N hydrochloric acid. K_b for hydroxylamine is 9.1×10^{-9}.

Since the base and acid are equal in concentration the volume at the endpoint will be double the original volume. Therefore, the concentration of the salt will be one-half the original concentration of the base.

Using Equation (7-61)

$$[OH^-] = \sqrt{\frac{K_w K_b}{C_s}}$$

$$= \sqrt{\frac{(1 \times 10^{-14})(9.1 \times 10^{-9})}{0.05}} = 4.3 \times 10^{-11}$$

$$= 10^{.63} \times 10^{-11} = 10^{-10.37}$$

$$pOH = 10.37$$

$$pH = 14 - 10.37$$

$$= 3.63$$

Using Equation (7-32)

$$[H^+] = \sqrt{K_a C_a}$$

$$K_a = \frac{K_w}{K_b} = \frac{1 \times 10^{-14}}{9.1 \times 10^{-9}} = 1.1 \times 10^{-6}$$

$$[H^+] = \sqrt{(1.1 \times 10^{-6})(0.05)} = 2.3 \times 10^{-4}$$

$$= 10^{.37} \times 10^{-4} = 10^{-3.63}$$

$$pH = 3.63$$

EXAMPLE 7-23: Calculate the percentage error in the titration of hydroxylamine if the titration is stopped at pH 7.00.

The solution of this problem is left to the student. The correct answer is an error of 91.6%.

Since hydroxylamine is a weaker base than acetic acid was an acid, the percentage error introduced by stopping a titration of hydroxylamine at pH 7.00 is considerably more serious than that experienced in Example 7-20.

Summary

The theory and examples expressed in this chapter can be summarized by learning the information listed in Table 7-2 and knowing how to apply it.

Table 7-2

Equations to be Used for Calculating pH Values of Solutions

Type of solution	Acid	Base
Strong electrolyte	a. $[H^+] = M$ or N	f. $[OH^-] = M$ or N
Weak electrolyte	b. $[H^+] = \sqrt{K_a C_a}$	g. $[OH^-] = \sqrt{K_b C_b}$
Weak electrolyte and its salt	c. $[H^+] = \dfrac{K_a C_a}{C_s}$	h. $[OH^-] = \dfrac{K_b C_b}{C_s}$
Salt of weak electrolyte	d. $[H^+] = \sqrt{\dfrac{K_w K_a}{C_s}}$	i. $[OH^-] = \sqrt{\dfrac{K_w K_b}{C_s}}$
	e. $[OH^-] = \sqrt{K_b C_b}$	j. $[H^+] = \sqrt{K_a C_a}$

Note that the left column in Table 7-2 lists the type of electrolyte solution and the middle and right columns the appropriate equation to be used to calculate the $[H^+]$ and $[OH^-]$ for acids and bases respectively. It should also be noted that all formulas for $[H^+]$ in systems containing acids are identical to those formulas for $[OH^-]$ in systems containing bases (only difference being K_a's for acids and K_b's for bases). Other points to be remembered are that all K's are in numerators, concentrations of the acid or base are always in the numerator and concentrations of the salt are always in the denominator.

It is advisable to learn and be able to use Table 7-2.

Questions

1. What is a conjugate acid-base pair?
2. Comment on the validity of the statement: "A neutral solution is a solution which has a pH of 7.00."
3. Show by appropriate calculations why the following statement is sometimes made: "The more dilute a solution of a weak acid becomes the stronger (i.e., the more ionized) it becomes."
4. Derive the general equation used to calculate the $[OH^-]$ in a solution of a weak base in water. Clearly define all terms and list any assumptions made.
5. Show with the aid of appropriate equations why a buffer composed of potassium formate (HCOOK) and formic acid (HCOOH) *resists* a change in pH upon the addition of (a) a small amount of HCl and (b) a small amount of NaOH.

6. Derive the general equation used to calculate the $[H^+]$ in a solution containing a weak acid and a salt of the weak acid.

7. A buffer solution containing a weak acid and its salt can be prepared by three different procedures. Describe each procedure.

8. Define buffer capacity.

9. Which of the following solutions would be most effective at resisting changes in pH with additions of a strong acid and/or a strong base? (a) $1 \times 10^{-4} N$ HCl, (b) $0.02 M$ $HC_2H_3O_2$ + $0.05 M$ $NaC_2H_3O_2$, or (c) $0.20 M$ $HC_2H_3O_2$ + $0.50 M$ NaOAc.

10. A buffer solution is composed of 3.0 mmoles HA and 1.0 mmoles NaA. Will this buffer be more effective at resisting a pH change with the addition of a small amount (up to 0.3 mmoles) of HCl or with the addition of a like amount of NaOH.

11. Derive the equation used for the calculation of $[H^+]$ in a solution containing the salt of a weak acid.

12. Derive an equation showing how the Brønsted-Lowry concept can be used to determine the $[OH^-]$ (and therefore the $[H^+]$) in a solution containing a salt of a weak acid.

Problems

Note: In all problems assume the temperature is 25°C unless otherwise specified.

1. Calculate the pH of solutions which have a hydrogen ion concentration of the following:
    (a) $1.7 \times 10^{-12} M$ (b) $4.6 \times 10^{-2} M$ (c) $3.3 \times 10^{-5} M$

2. Calculate the pH of a solution in which the hydroxide ion concentration is $3.9 \times 10^{-2} M$.

3. Convert the following pH values to $[H^+]$.
    (a) 7.2   (b) 3.8   (c) 11.6

4. At 50°C, K_w is equal to 6.30×10^{-14}. Calculate the pH of a neutral solution at 50°C.

5. Calculate the pH of a $0.1025 N$ solution of hydrochloric acid.

6. Calculate the pH of a solution prepared by dissolving 22.406 g of NaOH having a purity of 98.8% in water and dilution of the solution to 1000.0 ml.

7. Calculate the pH of a $0.15 M$ solution of formic acid ($K_a = 1.7 \times 10^{-4}$).

8. Calculate the pH of a $0.20 M$ solution of periodic acid ($K_a = 2.3 \times 10^{-2}$).

9. Calculate the pH of a $0.02 M$ solution of ethanolamine ($K_b = 2.8 \times 10^{-5}$).

10. Calculate the pH of a solution prepared by the addition of 13.2 g of sodium acetate to 100 ml of $0.10M$ acetic acid. Assume no change in volume.

11. Calculate the pH of a solution prepared by the addition of 25.0 ml of $0.100N$ NaOH to 50.0 ml of $0.250M$ acetic acid.

12. Calculate the pH of a solution prepared by the addition of 13.2 g of sodium acetate to 100 ml of $0.200N$ HCl.

13. In what molar ratio should acetic acid and sodium acetate be mixed in order to obtain a buffer having a pH of 4.00?

14. Calculate the buffer capacity of a solution containing the following:
(a) $0.100M$ formic acid and $0.100M$ sodium formate; K_a for formic acid $= 1.7 \times 10^{-4}$
(b) $0.250M$ formic acid and $0.250M$ sodium formate

15. Calculate the volume of $0.100N$ NaOH which must be mixed with 100 ml of $0.100M$ acetic acid to form a buffer with a pH of 4.25.

16. Calculate the pH of a solution prepared by the addition of 4.24 g of NH_4Cl to 100 ml of $0.1M$ aqueous ammonia solution.

17. Calculate the pH of a $0.100M$ solution of sodium formate.

18. Calculate the pH of a solution prepared by the addition of 50.0 ml of $0.100N$ NaOH to 25.0 ml of $0.200M$ hydrofluoric acid ($K_a = 6.8 \times 10^{-4}$).

19. Calculate the percentage error in the titration of formic acid ($K_a = 1.7 \times 10^{-4}$) if the titration is stopped at pH 5.2 using methyl red indicator.

20. Calculate the pH at the equivalence point in the titration of $0.100M$ ethanolamine ($K_b = 2.8 \times 10^{-5}$) with $0.100N$ HCl.

8

Acid-Base Titration Curves

The preceding chapter should make evident the fact that the pH of an acid solution changes in a predictable way when it is titrated with a base. The change in pH can be measured experimentally with a pH meter as the titration proceeds. The theory and operation of a pH meter will be described in Chapter 9.

Throughout most of an acid-base titration, the pH changes gradually as the titrant is added. As the equivalence point is approached, the rate of change in pH with volume increases. At the equivalence point the rate of change (slope) is at a maximum.

Strong Acid-Strong Base Titration

When a strong acid is titrated with a strong base, a very large and abrupt jump in pH occurs at the equivalence point.

> **EXAMPLE 8-1:** Construct the theoretical titration curve for the titration of 50.00 ml of $0.1000N$ HCl with $0.1000N$ NaOH.

165

Prior to the equivalence point in the titration, the hydrogen ion concentration is equal to the untitrated hydrochloric acid concentration left in the solution.

At the beginning of the titration

$$[H^+] = 0.1000N$$

$$pH = 1.00$$

After the addition of 10.00 ml NaOH, 1.000 millimole of the 5.000 millimoles of HCl has reacted with NaOH leaving 4.000 millimoles untitrated.

$$[H^+] = \frac{4.000 \text{ millimoles}}{60.00 \text{ ml}} = 6.67 \times 10^{-2}$$

$$pH = 1.18$$

After the addition of 25.00 ml NaOH, 2.500 millimoles HCl remains untitrated.

$$[H^+] = \frac{2.500 \text{ millimoles}}{75.00 \text{ ml}} = 3.33 \times 10^{-2}$$

$$pH = 1.48$$

After the addition of 40.00 ml NaOH, 1.000 millimole HCl remains untitrated.

$$[H^+] = \frac{1.000 \text{ millimole}}{90.00 \text{ ml}} = 1.11 \times 10^{-2}$$

$$pH = 1.95$$

After the addition of 49.00 ml NaOH, 0.100 millimoles HCl remains untitrated.

$$[H^+] = \frac{0.100 \text{ millimoles}}{99.00 \text{ ml}} = 1.01 \times 10^{-3}$$

$$pH = 3.00$$

After the addition of 49.90 ml NaOH, 0.010 millimoles HCl remains untitrated.

$$[H^+] = \frac{0.010 \text{ millimoles}}{99.90 \text{ ml}} = 1.00 \times 10^{-4}$$

$$pH = 4.00$$

After the addition of 49.99 ml NaOH, 0.0010 millimoles HCl remains untitrated.

$$[H^+] = \frac{0.0010 \text{ millimoles}}{99.99 \text{ ml}} = 1.00 \times 10^{-5}$$

$$pH = 5.00$$

At the equivalence point the pH of the solution at 25°C is 7.00 since the solution contains only water and a small amount of sodium chloride (which does not affect the pH). The further addition of sodium hydroxide serves merely as a dilution of the strong base and the pH is calculated accordingly.

After the addition of 50.01 ml NaOH, there is an excess of 0.0010 millimoles of NaOH.

$$[OH^-] = \frac{0.0010 \text{ millimoles}}{100.01 \text{ ml}} = 1.00 \times 10^{-5}$$

$$pH = 9.00$$

After the addition of 50.10 ml NaOH, an excess of 0.010 millimoles of NaOH exists.

$$[OH^-] = \frac{0.010 \text{ millimoles}}{100.10 \text{ ml}} = 1.00 \times 10^{-4}$$

$$pH = 10.00$$

After the addition of 51.00 ml NaOH, an excess of 0.100 millimoles of NaOH exists.

$$[OH^-] = \frac{0.100 \text{ millimoles}}{101.00 \text{ ml}} = 9.90 \times 10^{-4}$$

$$pH = 11.00$$

After the addition of 60.00 ml NaOH, an excess of 1.00 millimole of NaOH exists.

$$[\text{OH}^-] = \frac{1.000 \text{ millimole}}{110.00 \text{ ml}} = 9.09 \times 10^{-3}$$

$$\text{pH} = 11.96$$

After the addition of 75.00 ml NaOH, an excess of 2.500 milli-moles of NaOH exists.

$$[\text{OH}^-] = \frac{2.500 \text{ millimoles}}{125.00 \text{ ml}} = 2.00 \times 10^{-2}$$

$$\text{pH} = 12.30$$

After the addition of 100.00 ml NaOH, an excess of 5.000 milli-moles of NaOH exists.

$$[\text{OH}^-] = \frac{5.00 \text{ millimoles}}{150.00 \text{ ml}} = 3.33 \times 10^{-2}$$

$$\text{pH} = 12.52$$

These data have been plotted on Figure 8-1, along with similar data for other concentrations of acids and bases.

It is evident from Figure 8-1 that the more dilute the reacting solutions become, the less sharp the break at the end point.

The question might arise in the titration of a strong acid with a strong base whether the hydrogen ions contributed by the solvent water ever have a significant effect upon the pH. On the theoretical basis the answer is yes, although none of the points calculated above are affected by the ionization of water.

When 49.9999 ml of NaOH has been added, the concentration of unreacted HCl is $1 \times 10^{-7} N$. Assuming complete ionization $[\text{H}^+]$ from the unreacted HCl equals 1×10^{-7} and $[\text{Cl}^-]$ from the same source equals 1×10^{-7}.

Does this mean the pH is 7.00? No, the pH is 7.00 only at exactly 50.00 ml.

What is the actual pH at this point? To solve this problem the *Law of Electroneutrality* is applied. The Law of Electroneutrality merely states that the sum of the concentrations of all the cations in a solution multiplied by the charge on the cations is equal to the sum of the concentrations of all the anions in the solution multiplied by the absolute value of their charges.

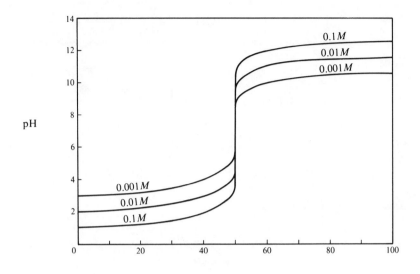

NaOH, ml

Figure 8-1

Titration curve for titration of hydrochloric acid
with sodium hydroxide at different concentrations.

Law of Electroneutrality:

$$\Sigma \, [i_c] \, Zi_c = \Sigma \, [i_a] \, Zi_a$$

where: $[i_c]$ = concentration of cations
         Zi_c = charge on cations
         $[i_a]$ = concentration of anions
         Zi_a = absolute value of charge on anions

The only cations in the solution in question are Na^+ and H^+. The
only anions in the solution are Cl^- and OH^-. Therefore $[Na^+]$ +
$[H^+]$ = $[Cl^-]_{titrated}$ + $[Cl^-]_{untitrated}$ + $[OH^-]$.

It can be seen that $[Na^+]$ = $[Cl^-]_{titrated}$ because they react mole for
mole.

Therefore,

$$[H^+] = [Cl^-]_{untitrated} + [OH^-]$$

$$[Cl^-]_{untitrated} = 1 \times 10^{-7}$$

$$[OH^-] = \frac{K_w}{[H^+]}$$

Thus,

$$[H^+] = 1 \times 10^{-7} + \frac{1 \times 10^{-14}}{[H^+]}$$

or

$$[H^+]^2 - 1 \times 10^{-7}[H^+] - 1 \times 10^{-14} = 0$$

Solving the equation yields

$$[H^+] = 1.62 \times 10^{-7}$$

$$pH = 6.79$$

The casual reader might think that the above problem is strictly theoretical and would never arise, but consider the following example:

EXAMPLE 8-2: Calculate the pH of a solution prepared by diluting 1.00 ml of $0.000100N$ HCl to 1.00 liter.

Applying the dilution law to this problem results in

$$V_1 N_1 = V_2 N_2$$

$$N_1 = \frac{V_2 N_2}{V_1}$$

$$= \frac{1.00 \text{ ml} \times 1.00 \times 10^{-4}N}{1.00 \times 10^3 \text{ ml}}$$

$$= 1.00 \times 10^{-7}N$$

If the contribution of $[H^+]$ from water is now ignored, the pH would be 7.00. It is impossible to neutralize an acid by dilution with water. Therefore, the $[H^+]$ from water must be considered and the problem is solved exactly as above and the correct pH is 6.79.

The pH of very dilute solutions of bases must also be solved in this manner.

Figure 8-2 shows a typical titration curve for the titration of NaOH

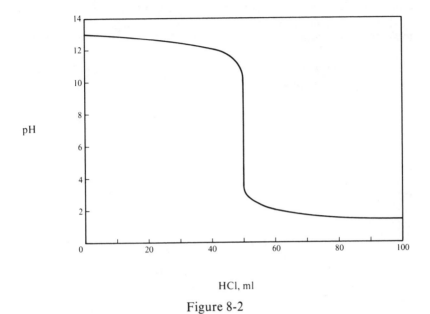

Figure 8-2

Titration curve for 0.100N NaOH with 0.100N HCl.

with HCl. It should be noted from either Figure 8-1 or Figure 8-2 that the selection of a visual indicator is not too critical in the titration of a strong acid with a strong base. Any indicator which changes color in the pH range represented by the nearly vertical line at the equivalence point will be satisfactory. The selection of a proper indicator to use in the titration of a weak acid or a weak base is much more critical as will be seen in the following section.

Weak Acid-Strong Base Titration

The calculation of the titration curve for a weak acid with a strong base involves several of the equations derived in Chapter 7 and tabulated in Table 7-2. Before any base is added, the solution has only the weak acid in it and Equation (b) is used. At all points after the start of the titration up to but not including the equivalence point a buffer exists since part of the acid has been converted to salt and part of it remains. Therefore, Equation (c) is used. At the theoretical end point, only the salt exists and Equation (d) or (e) is used. Past the end point the effect

172   *Acid-Base Titration Curves*

is only the dilution of the strong base in the solution since the salt of a weak acid has a negligible effect upon the pH of a strong base solution. Therefore, Equation (f) is used.

EXAMPLE 8-3:   Calculate the theoretical titration curve for the titration of 50.0 ml of 0.100M formic acid with 0.100N sodium hydroxide. K_a for HCOOH = 2.14 × 10^{-4}.

At the beginning of the titration, only HCOOH exists.

$$[H^+] = \sqrt{K_a C_a}$$
$$= \sqrt{(2.14 \times 10^{-4})(0.100)}$$
$$= 4.6 \times 10^{-3}$$
$$pH = 2.34$$

After 5.0 ml or 0.50 millimoles of sodium hydroxide have been added, 0.50 millimoles of sodium formate have formed and 4.50 millimoles of formic acid remain.

$$[H^+] = \frac{K_a C_a}{C_s} = \frac{2.14 \times 10^{-4} \times 4.50}{0.50}$$
$$= 1.93 \times 10^{-3}$$
$$pH = 2.71$$

After 25.0 ml NaOH have been added, the weak acid is one-half neutralized and the concentrations of acid form and salt form cancel. Therefore, $[H^+] = K_a$ and pH + pK_a.

$$[H^+] = K_a = 2.14 \times 10^{-4}$$
$$pH = 3.67$$

After 40.0 ml NaOH,

$$[H^+] = \frac{K_a C_a}{C_s} = \frac{2.14 \times 10^{-4} \times 1.00}{4.00}$$
$$= 5.35 \times 10^{-5}$$
$$pH = 4.27$$

After 49.0 ml NaOH,

$$[H^+] = \frac{2.14 \times 10^{-4} \times 0.10}{4.90}$$

$$= 4.4 \times 10^{-6}$$

$$pH = 5.36$$

After 49.9 ml NaOH,

$$[H^+] = \frac{2.14 \times 10^{-4} \times 0.01}{4.99}$$

$$= 4.3 \times 10^{-7}$$

$$pH = 6.37$$

After 50.0 ml NaOH have been added, the titration is at the equivalence point and only the salt of the acid exists. Since the volume of the solution has doubled, the concentration of sodium formate is $0.050M$.

$$[H^+] = \sqrt{\frac{K_w K_a}{C_s}}$$

$$= \sqrt{\frac{(1 \times 10^{-14})(2.14 \times 10^{-4})}{0.05}}$$

$$= 6.6 \times 10^{-9}$$

$$pH = 8.18$$

Past the equivalence point, the pH is controlled by the dilution effect of the sodium hydroxide and pH values are identical to those obtained in Example 8-1. These values are tabulated in Table 8-1 and the complete curve is shown in Figure 8-3.

Table 8-1

Volume of NaOH, ml	pH of solution
50.1	10.0
51.0	11.0
60.0	11.96
75.0	12.30
100.0	12.52

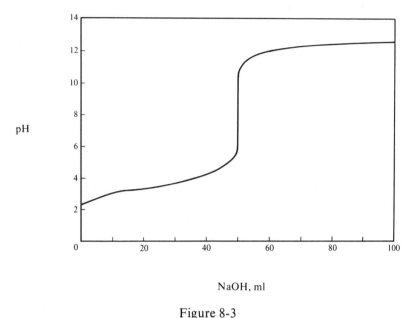

NaOH, ml

Figure 8-3

Titration curve for $0.100M$ formic acid
with $0.100N$ NaOH.

The extent of inflection at the equivalence point in the titration of a weak acid with a strong base depends upon two factors, K_a and the concentration of the reacting species. The higher the K_a (the stronger the acid) the greater the pH change at the equivalence point. This effect is shown in Figure 8-4. Data for Figure 8-4 are found in Table 8-2.

It can be seen from Figure 8-4 that a satisfactory titration of an acid with a pK_a of 9 or greater is not feasible in an aqueous solution. Table 8-2 indicates that with a K_a of 10^{-9} a change in pH of only 0.31 occurs from 1.0 ml before the equivalence point to 1.0 ml after the equivalence point. For a satisfactory titration of a $0.100M$ aqueous solution of a weak acid, the K_a must be at least as large as approximately 10^{-7}.

A lower concentration of base used in the titration means less inflection occurs at the end point. Consequently, there would be no change in the inflection with a change in C_a if the concentration of the titrant remained constant. However, to retain accuracy in the titration, the titrant must be correspondingly reduced in strength. Figure 8-5 shows curves for the titration of acetic acid solutions with sodium hydroxide both at $0.100M$, $0.0100M$, and $0.00100M$. Data for Figure 8-5 are found in Table 8-3.

Table 8-2

pH Values for Titration Curves of 0.100M Solutions of Weak Acids

Volume, NaOH ml	pK_a =	3	5	7	9
0.0		2.02	3.00	4.00	5.00
10.0		2.40	4.40	6.40	8.40
25.0		3.00	5.00	7.00	9.00
40.0		3.60	5.60	7.60	9.60
49.0		4.69	6.69	8.69	10.69
51.0		11.00	11.00	11.00	11.00
60.0		11.96	11.96	11.96	11.96
75.0		12.30	12.30	12.30	12.30

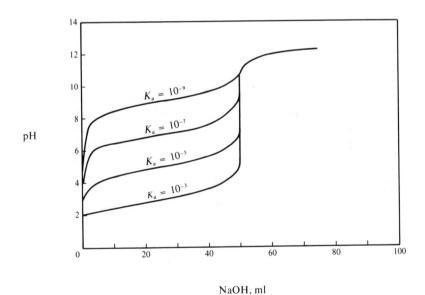

NaOH, ml

Figure 8-4

Titration curves for 0.100M solutions of
weak acids with different K_a's.

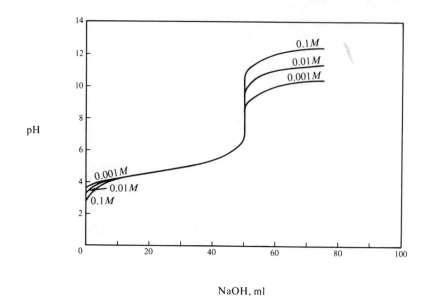

NaOH, ml

Figure 8-5

Titration curves for different concentrations
of acetic acid.

Table 8-3

pH Values for Titration Curves of Different Concentrations of
Acetic Acid with Equal Concentrations of Sodium Hydroxide

Volume, NaOH ml	[HOAc] = [NaOH]	0.1M	0.01M	0.001M
0.0		2.89	3.37	3.60
10.0		4.14	4.14	4.14
25.0		4.76	4.76	4.76
40.0		5.35	5.35	5.35
49.0		6.43	6.43	6.43
50.0		8.72	8.22	7.72
51.0		11.00	10.00	9.00
60.0		11.96	10.96	9.96
75.0		12.30	11.30	10.30

Two entries in Tables 8-2 and 8-3 need an explanation. The pH of a 0.10M solution of an acid with $K_a = 1 \times 10^{-3}$ is 2.02, but if it is calculated by the equation $[H^+] = \sqrt{K_a C_a}$ the pH would be 2.00. The pH of a 0.001M solution of acetic acid is 3.60, but if calculated with the above equation it would be 3.87. It should be recalled that a basic assumption in the derivation of that equation was that the hydrogen ion concentration was much less than the analytical concentration of the acid. When K_a is relatively high ($10^{-1} - 10^{-3}$) or in very dilute solutions, the hydrogen ion concentration may not be negligible compared to the analytical concentration of the acid. In that case the equation

$$[H^+]^2 = K_a(C_a - [H^+])$$

must be solved by use of the quadratic equation. This exact equation should be used whenever $[H^+]$ calculated by $\sqrt{K_a C_a}$ is approximately 5% or more of C_a.

Weak Base-Strong Acid Titration

The titration curves for the titration of weak bases with standard hydrochloric acid are identical to those represented in Figures 8-3, 8-4, and 8-5 except the slope is in the opposite direction.

Indicator Selection

An acid-base indicator is a weak organic acid or base which exhibits different colors in its ionized and nonionized forms. The conversion of the one color into the second color occurs at a definite pH which corresponds to the pK_a or pK_b for the substance. A common example is methyl red, a weak organic base of the azo type, which exhibits a red color in its acidic or ionized form and a yellow color in its basic or nonionized form. The conversion occurs at approximately pH 5.2. The equilibrium is expressed in Figure 8-6.

Another widely used indicator is phenolphthalein, a colorless dibasic acid (H_2Ind) which ionizes first to another colorless form ($HInd^-$) and then to a second ion ($Ind^=$) which is red. These equilibria are shown in Figure 8-7.

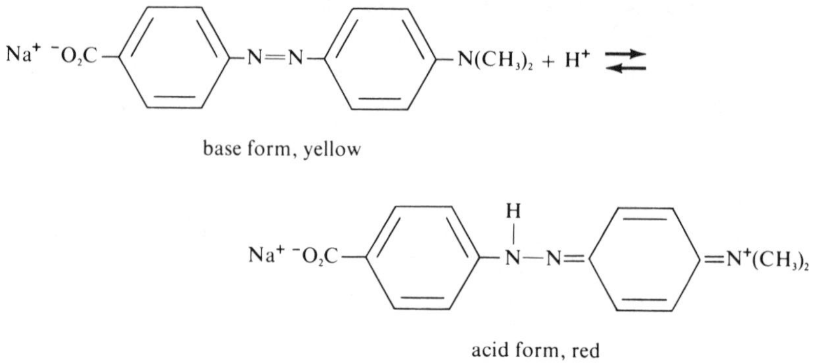

base form, yellow

acid form, red

Figure 8-6

Base (nonionized) and acid (ionized) forms
for methyl red.

An indicator of the weak acid form can be designated as HInd. The ionization of this acid is expressed as

$$HInd \rightleftharpoons H^+ + Ind^-$$

acid color   base color

When the indicator is in the nonionized form it exhibits the characteristic acid form and color and in the ionized form it exhibits the base form or base color. The ionization constant for the acid is

$$K_a = \frac{[H^+][Ind^-]}{[HInd]}$$

Rearrangement gives

$$[H^+] = \frac{K_a[HInd]}{[Ind^-]}$$

or

$$[H^+] = \frac{K_a[\text{acid color}]}{[\text{basic color}]}$$

It is obvious from the latter equation that the ratio of acid color to basic color for an indicator in solution depends upon the K_a for the indicator and the hydrogen ion concentration. For a particular indicator this ratio depends only upon the hydrogen ion concentration.

It is generally assumed that the human eye can detect a change in

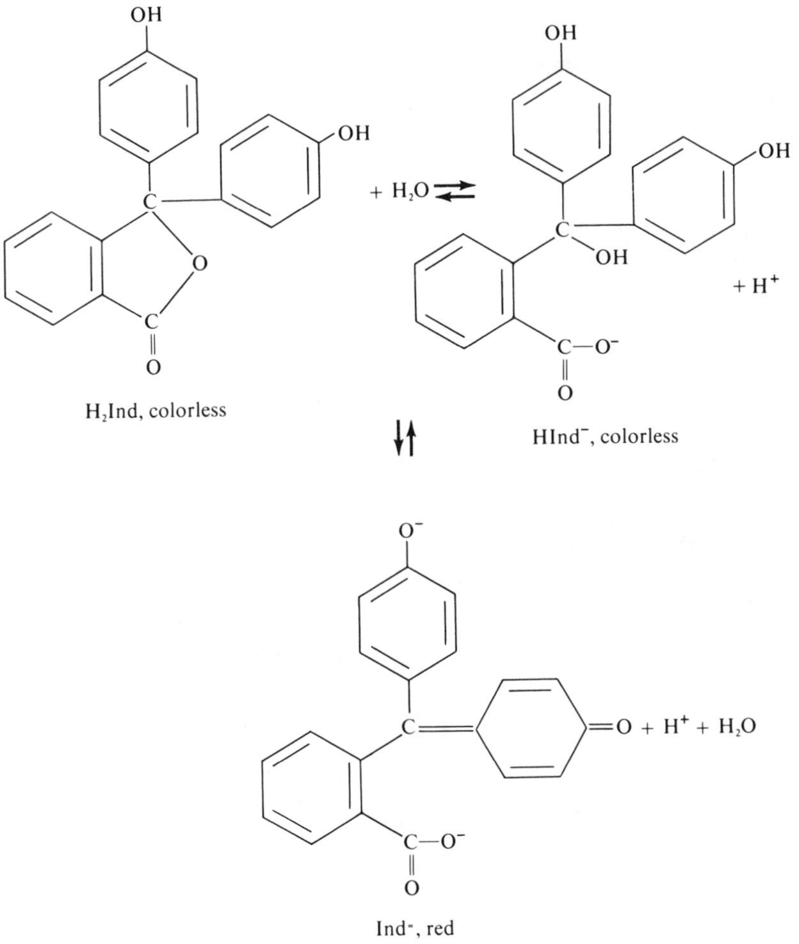

Figure 8-7

Acid-base forms for phenolphthalein.

color when one part of a second color is present with ten parts of the
first color. Thus, in the titration of an acid with sodium hydroxide
the first trace of a color change will occur when one part of base color
is present with ten parts of acid color. The last detectable change in
color will occur when there are ten parts of base color and one part of
acid color. Therefore, the following applies.

first base color: $\quad [H^+] = K_a\left(\dfrac{10}{1}\right)$

converting to pH: $pH = pK_a - 1$

last acid color: $[H^+] = K_a \left(\dfrac{1}{10}\right)$

converting to pH: $pH = pK_a + 1$

The interval in which these color changes occur $pK_a \pm 1$ is known as the pH range or pH interval for the indicator. A similar relationship can be easily derived for an indicator of the basic type.

An indicator should be selected which has a midpoint of pH range as near the actual equivalence point of a titration as is possible. Table 8-4 lists several acid-base indicators with their color changes and approximate range. Not all indicators have a range of 2 pH units because not all color combinations can be detected at the same ratios. Likewise, the range may not be symmetrical about pK because a higher ratio may be

Table 8-4

Some Acid-Base Indicators

Indicator	Color change		pH range
	Acid form	Base form	
Picric acid	Colorless	Yellow	0.1– 0.8
Crystal violet	Yellow	Blue	0.0– 1.8
Thymol blue	Red	Yellow	1.2– 2.8
	Yellow	Blue	8.0– 9.6
2,6–Dinitrophenol	Colorless	Yellow	2.0– 4.0
Bromphenol blue	Yellow	Blue	3.0– 4.6
Methyl orange	Red	Yellow	3.1– 4.4
Bromcresol green	Yellow	Blue	3.8– 5.4
Methyl red	Red	Yellow	4.2– 6.2
p–Nitrophenol	Colorless	Yellow	5.0– 7.0
Bromcresol purple	Yellow	Purple	5.2– 6.8
Bromthymol blue	Yellow	Blue	6.0– 7.6
Phenol red	Yellow	Red	6.8– 8.4
Cresol purple	Yellow	Purple	7.4– 9.0
	Red	Yellow	1.2– 2.8
Phenolphthalein	Colorless	Red	8.0– 9.6
Thymolphthalein	Colorless	Blue	9.3–10.5
Alizarin yellow	Yellow	Violet	10.1–12.0
2,4,6–Trinitrotoluene	Colorless	Orange	12.0–14.0

required for the observer to see one form than is required to see the other form.

Titrations of Mixtures of Acids

The pH of a solution is a measure of the hydrogen ion concentration regardless of the source of hydrogen ion. If a mixture of two strong acids, e.g., hydrochloric acid and nitric acid, is titrated with sodium hydroxide only a single inflection point occurs on the titration curve.

If a mixture containing a strong acid and a weak acid is titrated two inflection points occur. The stronger acid will be titrated first. The extent to which the titration of the first acid is completed before the second one begins to react is determined by the relative strengths of the two acids. Figure 8-8 shows the titration curve for a mixture of $0.05M$ HCl and $0.05M$ HOAc titrated with $0.1M$ NaOH.

A curve similar to Figure 8-8 will be obtained for a pair of weak acids if the K_a values for the two acids differ by approximately 10^4 or greater.

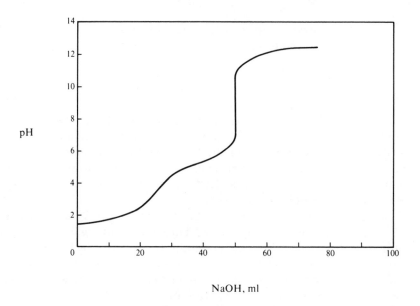

NaOH, ml

Figure 8-8

Titration curve for mixture of $0.05M$ HCl and
$0.05M$ HOAc with $0.1M$ NaOH.

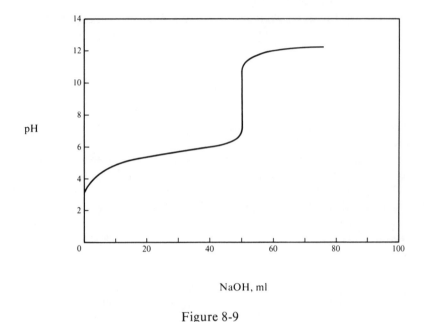

NaOH, ml

Figure 8-9

Titration curve for a mixture of acetic acid
and formic acid with 0.1M NaOH.

However, if the K_a values are much closer than this, the two acids cannot be differentiated and a single break occurs where the sum of the two
acids have been neutralized. Figure 8-9 shows the titration curve for a
mixture of acetic acid and formic acid. K_a for acetic acid is 1.8×10^{-5}
and for formic acid it is 1.7×10^{-4}. The ratio of K_a values is only
$10^{0.98}$. Therefore, only a single break is observed.

Sometimes a situation arises in a titration where the K_a values differ
more than that shown in Figure 8-9 but still are not sufficiently separated to allow quantitative evaluation of the results. Such a titration
can be used as qualitative evidence of two different acids and an approximate quantitative estimation may be possible. Figure 8-10 shows
the results of a titration of a mixture of 0.05M salicylic acid ($K_a =
1.05 \times 10^{-3}$) and 0.05$M$ potassium acid phthalate ($K_a = 3.9 \times 10^{-6}$).
Here the K_a values differ by a factor of 269 or $10^{2.43}$. The curve
definitely indicates that two acids are present although the first
equivalence point cannot be accurately determined from the curve.

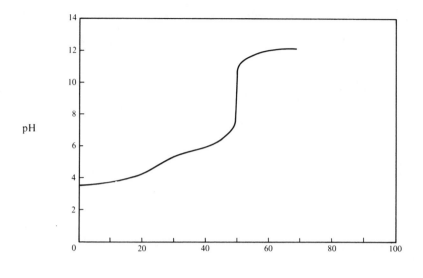

Figure 8-10

Titration curve for a mixture of 0.05 M salicylic acid
and 0.05 M potassium acid phthalate with 0.1 M
NaOH. (Small amount of ethyl alcohol added to
solubilize the salicylic acid.)

Polybasic Acids

Many organic acids have two or more acidic groups. These are known
as polybasic or polyprotic acids. Bases with more than one basic group
are known as polyacidic bases. Dibasic acids ionize in two stages as
follows:

$$H_2A \rightleftharpoons H^+ + HA^- \qquad \qquad \text{(8-1)}$$

$$HA^- \rightleftharpoons H^+ + A^= \qquad \qquad \text{(8-2)}$$

The ionization constants for the two steps may be written

$$K_1 = \frac{[H^+][HA^-]}{[H_2A]} \qquad \qquad \text{(8-3)}$$

$$K_2 = \frac{[H^+][A^=]}{[HA^-]} \qquad \qquad \text{(8-4)}$$

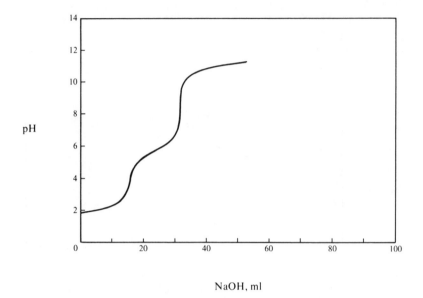

NaOH, ml

Figure 8-11

Titration curve for maleic acid, $K_1 = 1.2 \times 10^{-2}$
and $K_2 = 6.0 \times 10^{-7}$.

If the ratio K_1/K_2 is greater than approximately 10^4, a titration curve will show two distinct breaks. Figures 8-11, 8-12, and 8-13 show titrations for specific dibasic acids. In Figure 8-11, $(K_1/K_2) = 2 \times 10^4$ or $10^{4.30}$ and two sharp breaks occur. In Figure 8-12, $(K_1/K_2) = 636$ or $10^{2.80}$ and although two inflections points are apparent on the curve, the location of the first equivalence point cannot be determined with great certainty. In Figure 8-13, $(K_1/K_2) = 21$ or $10^{1.33}$ and no evidence of two equivalence points can be seen.

pH Values in Solutions of Polybasic Acids and Their Salts

A complete treatment of the equilibria involved in solutions of polybasic acids and their salts is complicated and beyond the scope of this text. However, in many cases, fairly valid assumptions can be made which simplify the treatment.

If the successive K_a values for a polybasic acid differ by at least 10^3 only adjacent species can coexist in any solution to an appreciable

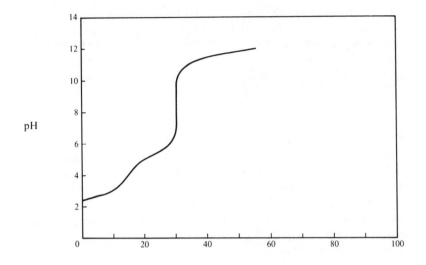

NaOH, ml

Figure 8-12

Titration curve for malonic acid, $K_1 = 1.4 \times 10^{-3}$ and $K_2 = 2.2 \times 10^{-6}$.

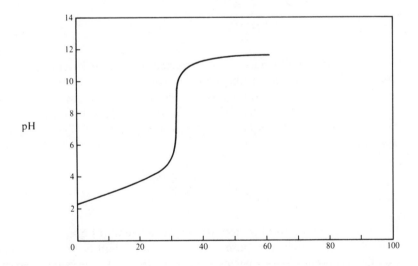

NaOH, ml

Figure 8-13

Titration curve for tartaric acid, $K_1 = 9.1 \times 10^{-4}$ $K_2 = 4.3 \times 10^{-5}$.

extent. Considering the dibasic acid H_2A the various possible forms in which it can exist are: H_2A, HA^-, and $A^=$. If K_1/K_2 is greater than 10^3 only H_2A plus HA^- or HA^- plus $A^=$ can coexist at appreciable concentrations. H_2A cannot coexist with $A^=$ because $A^=$ will act as a base and neutralize H_2A which acts as an acid as shown in Equation (8-5).

$$H_2A + A^= \rightleftharpoons 2HA^- \qquad (8\text{-}5)$$

The equilibrium in Equation (8-5) lies far to the right.

If the solution contains only the dibasic acid with no added salts, the hydrogen ion concentration is controlled only by the concentration of the acid and by K_1. Ionization of the acid occurs according to Equation (8-1) but since K_1 is much larger than K_2, ionization according to Equation (8-2) is repressed to a negligible extent. The hydrogen ion concentration is, therefore, given by Equation (7-32) which when applied to a solution of H_2A gives

$$[H^+] = \sqrt{K_1 C_a} \qquad (8\text{-}6)$$

Special care must be taken in the use of Equation (8-6) to make sure that $[H^+]$ calculated by its use does not exceed 5% of C_a. If it does, the exact expression, Equation (8-7), must be used.

$$[H^+]^2 = K_1(C_a - [H^+]) \qquad (8\text{-}7)$$

If the dibasic acid is partially neutralized, then at any point prior to the first equivalence point, the solution will contain a mixture of H_2A and HA^-. HA^- is the conjugate base of H_2A and can be considered as the salt form of the acid H_2A. Consequently, a buffer solution exists and the equation for a buffer solution [Equation (7-44)] can be applied.

$$[H^+] = \frac{K_1[H_2A]}{[HA^-]} \qquad (8\text{-}8)$$

Note that Equation (8-8) is merely a rearrangement of Equation (8-3).

At the first equivalence point the situation is slightly more complicated. The predominate species present at the first equivalence point is HA^-. HA^- can act as an acid, ionizing as in Equation (8-2).

$$HA^- \rightleftharpoons H^+ + A^= \qquad (8\text{-}2)$$

However, HA$^-$ also has basic properties and can react as in Equation (8-9).

$$HA^- + H^+ \rightleftharpoons H_2A \qquad \textbf{(8-9)}$$

Consequently [H$^+$] is not equal to [A$^=$] as would be indicated in Equation (8-2) because some of the [H$^+$] combines with HA$^-$ to form H$_2$A as indicated in Equation (8-9). Therefore, [A$^=$] is equal to the sum of the hydrogen ion remaining in solution plus that which reacted in Equation (8-9).

$$[A^=] = [H^+] + [H_2A] \qquad \textbf{(8-10)}$$

Rearranging gives

$$[H^+] = [A^=] - [H_2A] \qquad \textbf{(8-11)}$$

Unfortunately we do not know the magnitude of either component on the right-hand side of Equation (8-11). We can, however, express each component in terms of [HA$^-$] which we do know and an appropriate K_a. Thus, by rearranging Equation (8-3) we have

$$[H_2A] = \frac{[H^+][HA^-]}{K_1} \qquad \textbf{(8-12)}$$

and by rearranging Equation (8-4),

$$[A^=] = \frac{K_2[HA^-]}{[H^+]} \qquad \textbf{(8-13)}$$

Substituting Equations (8-12) and (8-13) into Equation (8-11) yields

$$[H^+] = \frac{K_2[HA^-]}{[H^+]} - \frac{[H^+][HA^-]}{K_1}$$

and

$$[H^+]^2 K_1 = K_1 K_2[HA^-] - [H^+]^2[HA^-]$$

Combining terms

$$[H^+]^2(K_1 + [HA^-]) = K_1 K_2[HA^-]$$

and

$$[H^+]^2 = \frac{K_1 K_2 [HA^-]}{K_1 + [HA^-]}$$

At normal concentrations $[HA^-]$ will be much larger than K_1 and the denominator will be approximately equal to $[HA^-]$. Therefore,

$$[H^+]^2 = \frac{K_1 K_2 [HA^-]}{[HA^-]}$$

$$= K_1 K_2$$

$$[H^+] = \sqrt{K_1 K_2} \tag{8-14}$$

Equation (8-14) can be used to calculate the pH at the first equivalence point and we find that the pH is independent of the concentration of HA^- unless the solution is very dilute in which case Equation (8-14) is not valid.

Equation (8-14) is used for a solution of HA^- regardless whether the solution results from the neutralization of H_2A to the first equivalence point, the acidification of $A^=$ to HA^-, or merely the dissolving of the salt $NaHA$ in water.

After the first equivalence point, but preceding the second equivalence point, another buffer solution results. Now the solution contains the weak acid HA^- and its conjugate base or salt, $A^=$. Equation (7-44) is applicable using the K which includes the two species present in the solution.

$$[H^+] = \frac{K_2 [HA^-]}{[A^=]} \tag{8-15}$$

Again note that Equation (8-15) is a rearranged form of Equation (8-4).

At the second equivalence point, the solution essentially contains only the species $A^=$. $A^=$, the salt of a weak acid or its conjugate base, undergoes hydrolysis resulting in a basic solution. The hydrogen ion concentration can be calculated from either Equation (7-55) or Equation (7-36).

$$[H^+] = \sqrt{\frac{K_w K_2}{C_s}} \tag{8-16}$$

or

$$[OH^-] = \sqrt{K_b C_b} \qquad (8\text{-}17)$$

where $K_b = (K_w/K_2)$ and $C_b = [A^=]$.

Past the second equivalence point, the net effect is the dilution of the titrant NaOH. The hydroxide concentration is equal to the molarity of the NaOH.

Titration Curve for Phosphoric Acid

Phosphoric acid H_3PO_4 is a tribasic acid, which ionizes stepwise as follows.

$$H_3PO_4 \rightleftharpoons H^+ + H_2PO_4^-$$

$$K_1 = \frac{[H^+][H_2PO_4^-]}{[H_3PO_4]} = 7.5 \times 10^{-3}$$

$$H_2PO_4^- \rightleftharpoons H^+ + HPO_4^=$$

$$K_2 = \frac{[H^+][HPO_4^=]}{[H_2PO_4^-]} = 6.2 \times 10^{-8}$$

$$HPO_4^= \rightleftharpoons H^+ + PO_4^{-3}$$

$$K_3 = \frac{[H^+][PO_4^{-3}]}{[HPO_4^=]} = 4.8 \times 10^{-13}$$

These K_a values are sufficiently separated that only adjacent species can coexist. However, only two inflection points are observed on the titration curve for phosphoric acid. The third hydrogen ion is too weak to be titrated in an aqueous solution.

> **EXAMPLE 8-4:** Calculate the theoretical titration curve for the titration of 25.0 ml of $0.100\,M$ H_3PO_4 with $0.100\,M$ NaOH.

0.0 ml:

$$[H^+]^2 = K_1(Ca - [H^+])$$
$$[H^+] = 2.39 \times 10^{-2}$$
$$pH = 1.62$$

12.5 ml:

$$[H^+] = \frac{K_1[H_3PO_4]}{[H_2PO_4^-]}$$

$$[H_3PO_4] = [H_2PO_4^-]$$

$$[H^+] = K_1 = 7.5 \times 10^{-3}$$

$$pH = 2.12$$

20.0 ml:

$$[H^+] = \frac{K_1[H_3PO_4]}{[H_2PO_4^-]}$$

$$[H_3PO_4] = \frac{5.0 \times 0.10}{45.0}$$

$$[H_2PO_4^-] = \frac{20.0 \times 0.10}{45.0}$$

$$[H^+] = 1.87 \times 10^{-3}$$

$$pH = 2.73$$

24.0 ml:

$$[H^+] = 3.13 \times 10^{-4}$$

$$pH = 3.51$$

25.0 ml:

$$[H^+] = \sqrt{K_1 K_2}$$

$$= \sqrt{(7.5 \times 10^{-3})(6.2 \times 10^{-8})}$$

$$= 2.16 \times 10^{-5}$$

$$pH = 4.67$$

26.0 ml:

$$[H^+] = \frac{K_2[H_2PO_4^-]}{[HPO_4^=]}$$

$$= 1.49 \times 10^{-6}$$

$$pH = 5.83$$

28.0 ml:

$$[H^+] = 4.55 \times 10^{-7}$$
$$pH = 6.34$$

30.0 ml:

$$[H^+] = 2.48 \times 10^{-7}$$
$$pH = 6.61$$

37.5 ml:

$$[H^+] = K_2 = 6.2 \times 10^{-8}$$
$$pH = 7.21$$

46.0 ml:

$$[H^+] = 1.18 \times 10^{-8}$$
$$pH = 7.93$$

48.0 ml:

$$[H^+] = 5.39 \times 10^{-9}$$
$$pH = 8.27$$

49.0 ml:

$$[H^+] = 2.58 \times 10^{-9}$$
$$pH = 8.59$$

50.0 ml:

$$[H^+] = \sqrt{K_2 K_3}$$
$$= \sqrt{(6.2 \times 10^{-8})(4.8 \times 10^{-13})}$$
$$= 1.73 \times 10^{-5}$$
$$pH = 9.76$$

51.0 ml:

$$[H^+] = \frac{K_3[HPO_4^=]}{[PO_4^{-3}]}$$

$$= 1.15 \times 10^{-11}$$

$$pH = 10.94$$

52.0 ml:

$$[H^+] = 5.52 \times 10^{-12}$$

$$pH = 11.26$$

55.0 ml:

$$[H^+] = 1.92 \times 10^{-12}$$

$$pH = 11.72$$

62.5 ml:

$$[H^+] = K_3 = 4.8 \times 10^{-13}$$

$$pH = 12.32$$

The data calculated in Example 8-4 are shown in Figure 8-14.

Phosphoric acid can be titrated as a tribasic acid by adding a large excess of calcium chloride after the second equivalence point is reached. This removes the phosphate ion leaving an equivalent amount of hydrochloric acid which can then be titrated with sodium hydroxide.

$$2Na_2HPO_4 + 3CaCl_2 \rightarrow Ca_3(PO_4)_2 + 4NaCl + 2HCl$$

Understanding the equilibria involved in solutions of phosphates enables the chemist to prepare buffer solutions of known pH. Consider the following example:

EXAMPLE 8-5: A phosphate buffer with a pH 7.00 is desired. (a) Calculate the molar ratio of phosphate salts to dissolve for pH 7.00. (b) Calculate the volume of $0.10M$ NaOH which must be added to 50.0 ml of $0.10M$ H_3PO_4 to give a pH 7.00.

Solution to part (a): Reference to Figure 8-14 shows that pH 7.00 occurs between the first and second equivalence points for

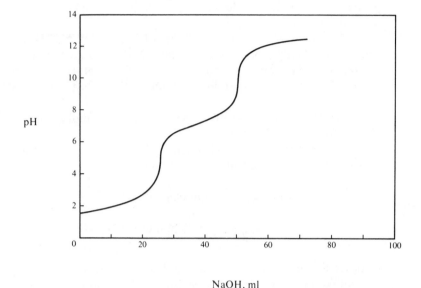

Figure 8-14

Titration curve for $0.10M$ H_3PO_4 with $0.10M$ NaOH.
$K_1 = 7.5 \times 10^{-3}$, $K_2 = 6.2 \times 10^{-8}$,
$K_3 = 4.8 \times 10^{-13}$.

phosphoric acid. The species present in solution is, therefore, the monohydrogen phosphate ion $HPO_4^=$ and the dihydrogen phosphate ion $H_2PO_4^-$. The hydrogen ion concentration is calculated with the use of the only ionization constant which includes both of these species, K_2. Therefore,

$$[H^+] = \frac{K_2[H_2PO_4^-]}{[HPO_4^=]}$$

$$\frac{1.00 \times 10^{-7}}{6.2 \times 10^{-8}} = \frac{[H_2PO_4^-]}{[HPO_4^=]}$$

$$\frac{[H_2PO_4^-]}{[HPO_4^=]} = 1.61$$

The pH 7.00 buffer solution should be prepared by mixing a ratio of 1.61 moles of NaH_2PO_4 with 1.00 moles of $NaHPO_4$ and diluting the mixture to a suitable volume. (The pH will be independent of the dilution volume; only the buffer capacity is affected by dilution.)

Solution to part (b):   The addition of 50.0 ml of NaOH will convert the H_3PO_4 quantitatively into NaH_2PO_4. Additional NaOH must be added to convert part of the NaH_2PO_4 into Na_2HPO_4. The total NaH_2PO_4 plus Na_2HPO_4 must equal 5.00 millimoles since both species originated from the 50.0 ml of $0.10M$ H_3PO_4. We know from part (a) that the molar ratio of NaH_2PO_4 to Na_2HPO_4 is 1.61.

Let X = millimoles of Na_2HPO_4 needed then $1.61X$ = millimoles of NaH_2PO_4 remaining and

$$X + 1.61X = 5.00 \text{ millimoles}$$

$$X = \frac{5.00}{2.61} = 1.92 \text{ millimoles } Na_2HPO_4$$

In order to form 1.92 millimoles of Na_2HPO_4 from NaH_2PO_4, 1.92 millimoles of NaOH must be added. The volume of NaOH which must be added to prepare the pH 7.00 buffer is 50.0 ml to the first equivalence point plus 19.2 ml to form the appropriate NaH_2PO_4/Na_2HPO_4 mixture. The correct answer to part (b) is, therefore, 69.2 ml.

Mixtures of NaOH, Na$_2$CO$_3$, and NaHCO$_3$

Sodium carbonate is the conjugate base of the very weak dibasic acid carbonic acid, H_2CO_3. The ionization constant expressions for carbonic acid are as follows:

$$H_2CO_3 \rightleftharpoons H^+ + HCO_3^-$$

$$K_1 = \frac{[H^+][HCO_3^-]}{[H_2CO_3]} = 4.3 \times 10^{-7}$$

$$HCO_3^- \rightleftharpoons H^+ + CO_3^=$$

$$K_2 = \frac{[H^+][CO_3^=]}{[HCO_3^-]} = 4.8 \times 10^{-11}$$

The constants for the ionization of the carbonate ion are calculated by use of the expression

$$K_b = \frac{K_w}{K_a}$$

Consequently the expressions can be written as follows:

$$CO_3^= + H_2O \rightleftharpoons HCO_3^- + OH^-$$

$$K_{b_1} = \frac{[HCO_3^-][OH^-]}{[CO_3^=]} = 2.1 \times 10^{-4}$$

$$HCO_3^- + H_2O \rightleftharpoons H_2CO_3 + OH^-$$

$$K_{b_2} = \frac{[H_2CO_3][OH^-]}{[HCO_3^-]} = 2.3 \times 10^{-8}$$

These K_b values are high enough that a titration of the carbonate ion with a strong acid will produce two breaks as shown in Figure 8-15. At the first equivalence point, carbonate has been converted to bi-

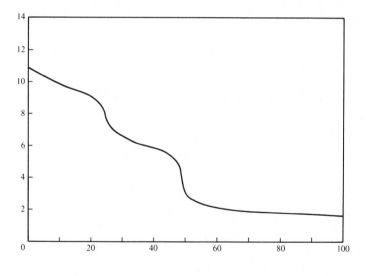

HCl, ml

Figure 8-15

Titration curve for Na$_2$CO$_3$ with 0.10M HCl.
$K_1 = 2.1 \times 10^{-4}$, $K_2 = 2.3 \times 10^{-8}$.

carbonate. This occurs at a pH of about 8.4. Phenolphthalein changes from pink to colorless at this point. The end point is not extremely sharp because the change in pH is gradual.

At the second equivalence point, the bicarbonate formed in the first step is converted to carbonic acid which decomposes into carbon dioxide and water:

$$H^+ + HCO_3^- \rightarrow H_2CO_3$$
$$H_2CO_3 \rightarrow CO_2 + H_2O$$

Much of the carbon dioxide is volatilized, thus causing a sharpening of the second end point. The pH at the second end point is about 4.0 and methyl orange changes from yellow to pink at this point.

When the only basic component of a sample is sodium carbonate, a titration with standard hydrochloric acid using phenolphthalein and methyl orange indicators yields two end points with the volume of acid reacted to the first end point equal to the volume of acid reacted between the two end points. When sodium carbonate is mixed with other basic components such as sodium hydroxide or sodium bicarbonate, the relationship between the volumes of acid to each end point is no longer equal.

Sodium hydroxide is a stronger base than sodium carbonate, however, an end point cannot be observed when the sodium hydroxide has been neutralized in a mixture of the two, because there is not a sufficient difference in their strengths. Figure 8-16 shows the titration curve for a mixture of sodium hydroxide and sodium carbonate. A greater volume of acid is required to reach the phenolphthalein end point than is required between the two end points. The volume of acid to reach the first end point represents the milliequivalents of sodium hydroxide plus the milliequivalents of sodium carbonate, whereas the volume of acid to reach the methyl orange end point represents the milliequivalents of sodium bicarbonate which resulted from the half neutralization at the first end point of the sodium carbonate originally present.

A sample containing a mixture of sodium carbonate and sodium bicarbonate will require less acid to reach the phenolphthalein end point than is required between the two end points. The volume of acid to reach the first end point corresponds to the milliequivalents of sodium carbonate being converted into sodium bicarbonate; whereas, the volume of acid consumed going on to the methyl orange end point represents the milliequivalents of sodium bicarbonate originally present *plus* the milliequivalents of sodium bicarbonate formed in reaching the first end point. A titration curve for a mixture of sodium carbonate and sodium bicarbonate is shown in Figure 8-17.

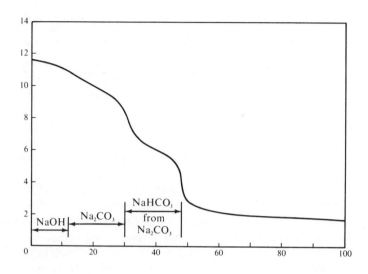

HCl, ml

Figure 8-16
Titration curve for a mixture of $NaOH$ and Na_2CO_3
with $0.10M$ HCl.

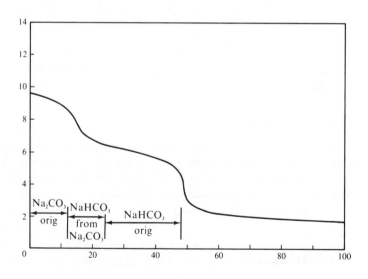

HCl, ml

Figure 8-17
Titration curve for a mixture of Na_2CO_3 and $NaHCO_3$
with $0.10M$ HCl.

A mixture of sodium hydroxide and sodium bicarbonate cannot exist. Sodium bicarbonate will act as an acid and neutralize the sodium hydroxide with perhaps some inert (neutral) impurities.

$$NaOH + NaHCO_3 \rightarrow Na_2CO_3 + H_2O$$

EXAMPLE 8-6: A 4.0000 g sample of impure sodium carbonate was dissolved in water and diluted to 250.0 ml in a volumetric flask. Using phenolphthalein as an indicator a 50.0 ml aliquot required 29.08 ml of 0.4025N HCl. Using methyl orange as an indicator a second 50.0 ml aliquot required 38.26 ml of the same acid. Calculate the percent composition of the sample.

Since the volume required for the methyl orange end point was less than twice the volume for the phenolphthalein end point, the sample must contain a mixture of sodium carbonate and sodium hydroxide with perhaps some inert (natural) impurities.

The 9.18 ml of acid required between the two end points represents the volume required to convert the sodium bicarbonate (formed from the sodium carbonate) into carbonic acid. The percent sodium carbonate is given by either of the following two expressions:

$$\%Na_2CO_3 = \frac{9.18\,\text{ml} \times 0.4025N \times \dfrac{Na_2CO_3}{1000} \times \dfrac{250}{50} \times 100}{4.0000\,\text{g}}$$

or

$$\%Na_2CO_3 = \frac{18.36\,\text{ml} \times 0.4025N \times \dfrac{Na_2CO_3}{2000} \times \dfrac{250}{50} \times 100}{4.0000\,\text{g}}$$

$$= 48.96\%$$

The 9.18 ml represents the volume of acid required to neutralize the sodium carbonate to the phenolphthalein end point and the 18.36 ml represents the volume of acid required to neutralize the sodium carbonate to the methyl orange end point.

The percent sodium hydroxide is given by the following:

$$\%NaOH = \frac{19.90\,\text{ml} \times 0.4025N \times \dfrac{NaOH}{1000} \times \dfrac{250}{50} \times 100}{4.0000\,\text{g}}$$

$$= 40.05\%$$

The sample also contained 10.99% inert impurities.

Analysis of Phosphate Mixtures

Mixtures containing phosphoric acid and its salts can be analyzed by a method similar to that for carbonate mixtures. The first end point in the titration with sodium hydroxide for the conversion of phosphoric acid, H_3PO_4, into dihydrogen phosphate, NaH_2PO_4, is indicated by the color change of methyl orange. The end point for the conversion of dihydrogen phosphate into monohydrogen phosphate, Na_2HPO_4, is indicated by the color change of phenolphthalein. The reverse titration with hydrochloric acid gives an end point for the conversion of trisodium phosphate, Na_3PO_4, into monohydrogen phosphate. This end point is indicated by the color change of phenolphthalein. At the methyl orange end point, the monohydrogen phosphate has been converted into dihydrogen phosphate. The mixture can be further complicated if we allow the presence of sodium hydroxide or hydrochloric acid in the sample. Figure 8-18 shows the relationships involved.

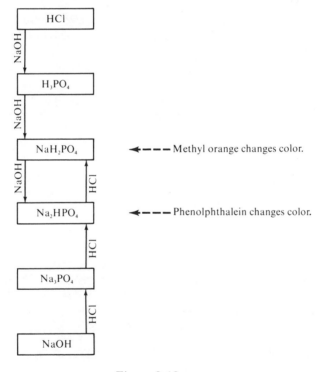

Figure 8-18

Titration of phosphate mixtures.

Only adjacent species shown in Figure 8-18 can coexist in solution. Any other combination will react with each other. Therefore, titrations with standard acid and standard base can be utilized to analyze the components of a mixture of the species. To improve the sharpness of the color change at the end point, the titrations should be performed at a slightly elevated temperature.

EXAMPLE 8-7: A 3.000 g sample containing a mixture of phosphate salts is titrated with 0.5000N hydrochloric acid requiring 28.40 ml of the acid to reach the methyl orange end point. A second 3.000 g portion of the sample required 11.00 ml of 0.6000 N sodium hydroxide to reach the phenolphthalein end point. Calculate the percentage composition of the sample.

The first titration indicates that the sample contains disodium hydrogen phosphate and/or sodium phosphate. The second titration indicates that the sample contains sodium dihydrogen phosphate and/or phosphoric acid. Since only adjacent species can coexist the presence of sodium phosphate and phosphoric acid may be ruled out. The sample, therefore, must contain a mixture of dihydrogen phosphate and monohydrogen phosphate.

$$\%Na_2HPO_4 = \frac{28.40\ ml \times 0.5000N \times \frac{Na_2HPO_4}{1000} \times 100}{3.000\ g}$$

$$= 67.2\%$$

$$\%NaH_2PO_4 = \frac{11.00\ ml \times 0.6000N \times \frac{NaH_2PO_4}{1000} \times 100}{3.000\ g}$$

$$= 26.4\%$$

The relationships between the titration volumes and the qualitative composition of the mixture is shown in Table 8-5. *A* represents the volume of standard base required to reach the methyl orange end point. *B* represents the *additional* volume of standard base required to reach the phenolphthalein end point. *C* represents the volume of standard acid required to reach the phenolphthalein end point. *D* represents the *additional* volume of standard acid required to reach the methyl orange end point.

Table 8-5

Relationships Between Titration Volumes for Phosphate Mixtures

Species present	Volume NaOH to m.o. end pt	Additional volume NaOH to ph. end pt	Volume HCl to ph. end pt	Additional volume HCl to m.o. end pt
HCl	*A*	0	0	0
H_3PO_4	*A*	*B* = *A*	0	0
$H_2PO_4^-$	0	*B*	0	0
HPO_4^{--}	0	0	0	*D*
PO_4^{---}	0	0	*C*	*D* = *C*
OH^-	0	0	*C*	0
$HCl + H_3PO_4$	*A*	*B* < *A*	0	0
$H_3PO_4 + H_2PO_4^-$	*A*	*B* > *A*	0	0
$H_2PO_4^- + HPO_4^{--}$	0	*B*	0	*D*
$HPO_4^{--} + PO_4^{---}$	0	0	*C*	*D* > *C*
$PO_4^{---} + OH^-$	0	0	*C*	*D* < *C*

Questions

1. The sharpness of the inflection of the end point in the titration of a weak acid with a strong base depends upon two factors. Discuss in a qualitative manner how these factors affect the sharpness.

2. Sketch the expected titration curve for the following solutions. In each case the titration is performed using $0.10N$ NaOH. You need not plot actual pH values but you should plot volumes accurately.
   (a) 50.0 ml of a $0.10M$ solution of a tribasic acid having $K_1 = 2.1 \times 10^{-2}$, $K_2 = 2.0 \times 10^{-7}$, $K_3 = 4.6 \times 10^{-13}$
   (b) A solution prepared by mixing 20.0 ml of $0.10M$ benzoic acid ($K_a = 6.3 \times 10^{-5}$) with 25.0 ml of $0.10N$ HCl.
   (c) A solution prepared by mixing 15.0 ml of $0.10M$ hydrofluoric acid ($K_a = 6.7 \times 10^{-4}$) and 35.0 ml of $0.1M$ formic acid ($K_a = 1.8 \times 10^{-4}$).
   (d) 50.0 ml of a $0.10M$ solution of oxalic acid having $K_1 = 6.5 \times 10^{-2}$, $K_2 = 6.1 \times 10^{-5}$.
   (e) 20.0 ml of a $0.10M$ solution of a tribasic acid having $K_1 = 1.2 \times 10^{-3}$, $K_2 = 8.4 \times 10^{-4}$, $K_3 = 2.7 \times 10^{-8}$.

3. It is often said that most indicators have a transition range of 2 pH units. Explain with the aid of appropriate equations why this statement is generally true. Also comment on why not all indicators have a pH range of 2 and why some ranges are not symmetrical about the end point.

4. What factor must be considered to determine the proper indicator for an acid-base titration?

5. What is the approximate pK of the weakest acid or base which can be titrated in water?

6. What must be the difference in the strengths of two acids in order to differentiate between them in a titration?

7. Write equations showing why only adjacent species of the series HCl, H_3PO_4, $H_2PO_4^-$, HPO_4^-, PO_4^{-3} can coexist.

Problems

1. Calculate the pH of a solution prepared by the addition of 25.0 ml of 0.100N NaOH to 50.0 ml of 0.100N HCl.

2. Calculate the pH of a solution prepared by the addition of 75.00 ml of 0.100N NaOH to 50.0 ml of 0.100N HCl.

3. Calculate the pH of a solution prepared by the dilution of 1.00 ml of $1 \times 10^{-5}N$ HCl to 1.00 liter.

4. Calculate the theoretical titration curve for the titration of 50.0 ml of 0.100M benzoic acid with 0.100N NaOH. K_a for benzoic acid = 6.3×10^{-5}.

5. Calculate the theoretical titration curve for a 0.100M solution of ethanolamine with 0.100M HCl. K_b for ethanolamine = 2.8×10^{-5}.

6. Hydrazine is a weak base with a K_b of 1.3×10^{-6}. Show by proper calculations which of the following indicators would be most acceptable for the titration of 0.02M hydrazine with 0.02N HCl.

Indicator	pH range
Phenolphthalein	8.0–9.6
Bromthymol blue	6.0–7.6
Methyl red	4.2–6.2
Methyl orange	3.1–4.4
Paramethyl red	1.0–3.0

7. Calculate the pH of a 100 ml solution containing 4.5 mmoles of HCl and 4.5 mmoles of HOAc.

8. O-phthalic acid is a weak dibasic acid with $K_1 = 1.2 \times 10^{-3}$ and $K_2 = 3.9 \times 10^{-6}$. A 40.0 ml sample of a 0.20M solution of the acid is titrated with 0.20M NaOH.
   (a) Calculate the pH at the start of the titration.
   (b) Calculate the pH at the first equivalence point.
   (c) Calculate the pH after the addition of 48.2 ml of the 0.20M NaOH.
   (d) Calculate the pH at the second equivalence point.
   (e) Calculate the pH after the addition of 90.0 ml of the 0.20M NaOH.

9. Explain with the aid of appropriate equations why many indicators have a pH range of approximately 2 pH units.

10. You have been given the task of preparing pure potassium acid phthalate, $KHC_8H_4O_4$, by the Wilson Chemical Co. Available for your use is O-phthalic acid, $H_2C_8H_4O_4$, and standard $0.1000N$ KOH. The method calls for the addition of KOH until the pH reaches a certain value. Calculate the correct pH value to reach. The ionization constants for O-phthalic acid are $K_1 = 1.2 \times 10^{-3}$ and $K_2 = 3.9 \times 10^{-6}$.

11. Calculate the pH at the beginning and at the end point in the titration of a 50.0 ml sample of $0.100M$ potassium acid oxalate (KHC_2O_4) with $0.100M$ NaOH. The ionization constants for oxalic acid are $K_1 = 8.8 \times 10^{-2}$ and $K_2 = 5.1 \times 10^{-5}$.

12. Phosphoric acid is a weak tribasic acid with $K_1 = 7.5 \times 10^{-3}$, $K_2 = 6.2 \times 10^{-8}$, and $K_3 = 4.8 \times 10^{-13}$. A 25.0 ml sample of $0.200M$ H_3PO_4 was titrated with $0.200M$ NaOH.
    (a) Calculate the pH at the start of the titration.
    (b) Calculate the pH after the addition of 15.0 ml of NaOH.
    (c) Calculate the pH after the addition of 25.0 ml of NaOH.
    (d) Calculate the pH after the addition of 35.0 ml of NaOH.
    (e) Calculate the pH after the addition of 50.0 ml of NaOH.
    (f) Calculate the pH after the addition of 62.5 ml of NaOH.

13. Calculate the volume of $0.100M$ NaOH which must be added to 50.0 ml of $0.100M$ H_3PO_4 to give a pH value of 7.2.

14. A chemist received different mixtures for analysis with the statement that they contained either NaOH, $NaHCO_3$, Na_2CO_3, or possible mixtures of these substances with inert material. From the following titration data, qualitatively identify the samples.
    Sample 1. With phenolphthalein as an indicator, 24.32 ml of acid was required. A *duplicate* sample required 48.64 ml of acid with methyl orange as indicator.
    Sample 2. The addition of phenolphthalein caused no change in color. With methyl orange, 38.47 ml of the acid was required.
    Sample 3. Using phenolphthalein 15.29 ml of acid was used. The addition of methyl orange required an *additional* 27.85 ml of acid.
    Sample 4. The sample when titrated using phenolphthalein required 38.48 ml of acid. When methyl orange was added the solution showed its acidic color with no additional acid.

15. A 1.0000 g sample containing Na_2CO_3 and either NaOH or $NaHCO_3$ was dissolved in water and titrated with $0.3800N$ HCl requiring 22.48 ml to reach the phenolphthalein end point and an additional 24.25 ml to reach the methyl orange end point. Calculate the percentage of each basic substance in the sample.

16. A 1.0000 g sample as above was titrated with $0.3800N$ HCl requiring 30.88 ml to reach the phenolphthalein end point and a total of 49.20 ml to

reach the methyl orange end point.   Calculate the percentage of each
basic substance in the sample.

17.   A 3.000 g sample was known to contain $Na_3PO_4 \cdot 12H_2O$, or $Na_2HPO_4 \cdot$
      $12H_2O$, or $NaH_2PO_4 \cdot H_2O$, or a mixture of these substances with inert
      material.   When the sample is titrated with $0.500N$ HCl, using methyl
      orange indicator, 14.0 ml of acid is required.   A second 3.000 g sample
      requires 5.00 ml of $0.600N$ NaOH using phenolphthalein as an indicator.
      Calculate the percentage composition of the sample.

18.   A liquid sample is known to contain any possible combination of species
      shown in Table 8-5.   A 50.0 ml portion of the sample requires 28.2 ml of
      $0.458N$ NaOH to reach the methyl orange end point and an additional
      18.4 ml of the base to reach the phenolphthalein end point.   Calculate the
      molarity of each active ingredient in the sample.

9

Theory of Oxidation-Reduction Reactions and Titrations

Basic Principles

Oxidation-reduction reactions are those in which there is a net change in oxidation number of reacting species. Oxidation is an increase in oxidation number of a species resulting from a loss of electrons. Reduction is a decrease in oxidation number of a species resulting from a gain of electrons. Oxidation cannot occur in a reaction unless reduction also occurs. The term *redox* is often used to describe the process of gain and loss of electrons.

An oxidizing agent is a substance which causes another species to be oxidized. It in turn gains electrons and becomes reduced. A general equation for this process is

$$\text{oxidized form} + ne^- = \text{reduced form}$$

where n is the number of electrons gained.

A reducing agent is a substance which causes another species to be reduced. It in turn loses electrons and becomes oxidized. A general equation for this process is

$$\text{reduced form} = \text{oxidized form} + ne^-$$

A redox reaction, therefore, is a reaction involving a transfer of electrons from a reducing agent to an oxidizing agent. Each reaction written above is called a *half reaction* and a redox reaction is the sum of two half reactions. The two species involved in a half reaction are called a couple. The total number of electrons gained by the oxidizing agent in a redox reaction must equal the total number of electrons lost by the reducing agent.

Balancing Redox Equations

The balancing of redox equations gives many students trouble. However, if a systematic method is followed, redox equations can be quickly balanced. A set of rules is given which if closely followed will enable the student to balance redox equations.

1. Determine the oxidation number of all elements involved. Hydrogen in all its compounds is always $+1$ except in metal hydrides (-1) and oxygen in all its compounds is always -2 except in peroxides (-1) and superoxides $(-1/2)$. The sum of the oxidation numbers of all elements in a neutral compound is zero. The sum of the oxidation numbers in an ionic species is equal to the charge on the ion. The oxidation number of any substance in its elemental form is zero.

2. Determine the change in oxidation states. In complex situations attribute all the oxidation to one specific element and all the reduction to one specific element (usually different).

3. Balance the changes in oxidation states. To do this use the number of electrons gained per molecule of the oxidizing agent as the coefficient for the reducing agent and use the number of electrons lost per molecule of the reducing agent as the coefficient for the oxidizing agent.

4. Balance all other constituents by inspection. This step can be accomplished without accounting for one element (usually oxygen or hydrogen).

5. *Check* by counting the number of atoms of the omitted element from step 4 on each side of the equation. If there is an equal number of atoms on each side of the equation, it is probably correctly

balanced. If the number of such atoms is not equal, the equation is definitely incorrectly balanced.

The use of the above rules is illustrated in the following examples.

$$\overset{+1}{Na_2}\overset{+3}{C_2}\overset{-2}{O_4} + \overset{+1}{K}\overset{+7}{Mn}\overset{-2}{O_4} + \overset{+1}{H_2}\overset{+6}{S}\overset{-2}{O_4}$$

$$\rightarrow \overset{+4-2}{CO_2} + \overset{+1}{H_2}\overset{-2}{O} + \overset{+1}{Na_2}\overset{+6-2}{SO_4} + \overset{+2'+6-2}{MnSO_4} + \overset{+1+6-2}{K_2SO_4}$$

The oxidation number for each element is indicated by the small numbers above the formula of each substance. It is seen that each carbon atom in sodium oxalate loses one electron and the manganese atom in potassium permanganate gains five electrons. Balancing errors are avoided by connecting the oxidized and reduced form of each couple with a tie-line as follows:

$$Na_2C_2O_4 + KMnO_4 + H_2SO_4$$

$$+5e^-$$

$$\rightarrow CO_2 + H_2O + Na_2SO_4 + MnSO_4 + K_2SO_4$$

$$-1e^- \times 2 = -2e^-$$

Applying rule 3 gives a coefficient of 5 for sodium oxalate and 10 for carbon dioxide and also 2 for potassium permanganate and manganous sulfate. This gives

$$5Na_2C_2O_4 + 2KMnO_4 + H_2SO_4$$

$$\rightarrow 10CO_2 + H_2O + Na_2SO_4 + 2MnSO_4 + K_2SO_4$$

Coefficients for the other species are determined by inspection. The only source of sodium in the reactants is sodium oxalate and all of it must end up as sodium sulfate. The coefficient for Na_2SO_4 is, therefore, 5. The only source of potassium in the reactants is potassium permanganate and it must end up as potassium sulfate. The coefficient for potassium sulfate must, therefore, be 1. The sulfate radical occurs in three products, namely as sodium, manganous, and potassium sulfates. All the sulfate must have come from the sulfuric acid. The coefficient for sulfuric acid is, therefore, 8. All the hydrogen originating in the sulfuric acid ends up in the water formed by the reaction. The coefficient of water is also 8. Thus, a coefficient has been determined for each species. The equation is, therefore, balanced.

$$5Na_2C_2O_4 + 2KMnO_4 + 8H_2SO_4$$
$$\rightarrow 10CO_2 + 8H_2O + 5Na_2SO_4 + 2MnSO_4 + K_2SO_4$$

However, rule 5 should be applied. One notes that in the balancing of the equation no concern was paid to the number of atoms of oxygen on either side of the equation. If we now count the oxygen atoms, it is seen that there are 60 on each side of the equation. Thus, the equation is correctly balanced.

In many reactions only a portion of the atoms of a certain species undergo oxidation or reduction. This is particularly true when nitric acid or sulfuric acid is used as an oxidizing agent. For example,

$$CuS + HNO_3 \rightarrow Cu(NO_3)_2 + S + NO + H_2O$$

The oxidation changes are determined and the tie lines are drawn.

$$\overset{+3e^-}{\overbrace{}}$$

$$\overset{-2}{C}uS + \overset{+5}{H}NO_3 \rightarrow Cu(NO_3)_2 + \overset{0}{S} + \overset{+2}{N}O + H_2O$$

$$\underset{-2e^-}{\underbrace{}}$$

The correct coefficient for cupric sulfide and sulfur is 3 and the correct coefficient for nitric oxide is 2. However, the coefficient for nitric acid is not 2 because not all the nitrogen is being changed from $+5$ to $+2$; some of it remains at $+5$ in cupric nitrate. Only three of the four species connected by tie lines can be determined directly from electron transfer. This gives

$$3CuS + HNO_3 \rightarrow Cu(NO_3)_2 + 3S + 2NO + H_2O$$

Obviously, if there are 3 cupric sulfides there must be 3 cupric nitrates. Thus, in addition to the two atoms of nitrogen which underwent reduction, there were 6 atoms which remained unchanged. The coefficient for nitric acid is, therefore, 8. All hydrogen must end up in the product water, so the coefficient for water is 4. Thus, the balanced equation is

$$3CuS + 8HNO_3 \rightarrow 3Cu(NO_3)_2 + 3S + 2NO + 4H_2O$$

Checking the equation finds 24 atoms of oxygen on each side.

A reaction in which both the oxidizing and reducing agents yield the same product sometimes arises. The determination of manganous salts by titration with potassium permanganate in a neutral solution is an example.

$$
\begin{array}{c}
\overset{+3e^-}{\overbrace{\hspace{3cm}}} \\
\overset{+2}{MnCl_2} + \overset{+5}{KMnO_4} \rightarrow \overset{+4}{MnO_2} + KCl + ZnCl_2 \\
\underset{-2e^-}{\underbrace{\hspace{4cm}}}
\end{array}
$$

Two molecules of potassium permanganate are reduced while 3 molecules of manganous chloride are oxidized. Both species form manganese dioxide giving a total of 5 molecules. The complete balanced equation is

$$3MnCl_2 + 2ZnO + 2KMnO_4 \rightarrow 5MnO_2 + 2KCl + 2ZnCl_2$$

A similar type reaction is known as autooxidation-reduction where a single species is both oxidized and reduced.

$$
\begin{array}{c}
\overset{+2e^-}{\overbrace{\hspace{3cm}}} \\
\overset{+6}{ReF_6} + SiO_2 \rightarrow \overset{+4}{ReF_4} + \overset{+7}{ReO_3F} \\
\underset{-1e^-}{\underbrace{\hspace{4cm}}}
\end{array}
$$

The correctly balanced equation is

$$3ReF_6 + 3SiO_2 \rightarrow ReF_4 + 2ReO_3F + 3SiF_4$$

Often an oxidation-reduction equation is written in a net ionic form. The dissolution of copper in nitric acid is an example of this type of equation.

$$
\begin{array}{c}
\overset{+1e^-}{\overbrace{\hspace{3cm}}} \\
\overset{0}{Cu} + H^+ + \overset{+5}{NO_3^-} \rightarrow \overset{+2}{Cu^{++}} + \overset{+4}{NO_2} + H_2O \\
\underset{-2e^-}{\underbrace{\hspace{4cm}}}
\end{array}
$$

Balancing just the oxidation-reduction portion of the equation yields

$$1Cu + H^+ + 2NO_3^- \rightarrow 1Cu^{++} + 2NO_2 + H_2O$$

It is seen that the total charge on the right-hand side must be +2 because the only ionic species is the copper(II) ion and there is one of them. The net ionic charge on the left-hand side must also be +2. Since there is a −2 for the two nitrate ions there must be a total of four hydrogen ions present in order to give a net charge of +2 for the left-hand side of the equation. The four hydrogen ions provide for two molecules of water and the complete balanced equation is

$$Cu + 4H^+ + 2NO_3^- \rightarrow Cu^{++} + 2NO_2 + 2H_2O$$

In all the previous examples only one species was being oxidized and one species was being reduced. When more than one species is being oxidized or reduced, the situation becomes more complicated. However, rule 2 says that in complex situations assume that all the oxidation is due to one species and that all the reduction is due to one species. This principle is illustrated in the following equation:

$$\overset{0}{As_2S_3} + \overset{}{Cl_2} + H_2O \rightarrow \overset{+5}{H_3}\overset{+6}{AsO_4} + \overset{-1}{H_2SO_4} + HCl$$

One can readily see that chlorine is reduced from an oxidation state of 0 to −1. What is being oxidized? Actually both arsenic and sulfur are being oxidized simultaneously. However since the ratio of atoms of the two elements must remain constant, it is possible to attribute all of the change to either of the two elements. Although this method gives unrealistic oxidation numbers, the equation does balance correctly with less likelihood of making an error.

First assume all the change is due to arsenic.

$$\overset{+1e^- \times 2 = +2e^- + 1e^-}{\overline{\underset{As_2S_3 + Cl_2 + H_2O \rightarrow H_3AsO_4 + H_2SO_4 + HCl}{\underset{-9 \quad\quad 0 \quad\quad\quad\quad +5 \quad\quad +6 \quad\quad -1}{}}}}$$

$$\underline{-14e^- \times 2 = -28e^- - 14e^-}$$

The sulfur on the right-hand side has an oxidation number of +6. If we assume it to be also +6 in As_2S_3, then the "oxidation number" for arsenic becomes −9. Note that there is a common factor in the numbers of electrons gained and lost. This factor of 2 is taken out before any coefficients are assigned. The balanced equation is

$$As_2S_3 + 14Cl_2 + 20H_2O \rightarrow 2H_3AsO_4 + 3H_2SO_4 + 28HCl$$

What if we assume the change to be due only to sulfur?

$$+1e^- \times 2 = +2e^- + 1e^-$$

$$\overset{-3\frac{1}{3}}{As_2S_3} + \overset{0}{Cl_2} + H_2O \rightarrow \overset{+5}{H_3AsO_4} + \overset{+6}{H_2SO_4} + \overset{-1}{HCl}$$

$$-9\frac{1}{3} \times 3 = -28e^- - 14e^-$$

Although the assigned "oxidation number" for sulfur is $-3\frac{1}{3}$, we see that the total loss of electrons per molecule is again 28 and the equation is balanced as shown before.

The above method of attributing all the change to a single element is extremely useful in balancing organic reactions. For organic compounds always assign hydrogen $+1$, oxygen -2, and attribute all the change to carbon. When nitrogen, sulfur, or other elements are also present in the organic compound, it is advisable to assume that they also do not undergo a change in oxidation number.

If the five rules given above are closely followed, the balancing of redox equations can readily be mastered.

Redox equations may also be balanced by using the half reaction method. Again a set of rules may be developed which if followed closely will enable the student to accurately balance the equation.

1. Write skeleton half reactions involving the essential species that undergo oxidation and reduction.
2. Balance the atoms in each half reaction, adding H^+ or H_2O as necessary.
3. Balance the charge for each half reaction by adding electrons to either the left or right side of the equation. If electrons appear on the left side, the half reaction is a reduction and if electrons appear on the right side, the reaction is an oxidation.
4. Multiply each half reaction by a coefficient such that the number of electrons involved in each half reaction is the same.
5. Combine the two half reactions and balance all atoms not directly involved in the oxidation or reduction by inspection.
6. Check by counting the number of each type of atoms on both sides of the equation and make sure that the net charge is the same on each side.

The use of these rules is illustrated by the following equation balanced previously by the other method.

$$Na_2C_2O_4 + KMnO_4 + H_2SO_4$$
$$\rightarrow CO_2 + H_2O + Na_2SO_4 + MnSO_4 + K_2SO_4$$

The two half reactions are:

$$Na_2C_2O_4 \rightarrow CO_2$$

and

$$KMnO_4 \rightarrow MnSO_4$$

Balance the atoms in each half reaction.

$$Na_2C_2O_4 \rightarrow 2CO_2 + 2Na^+$$
$$KMnO_4 + 8H^+ + SO_4^= \rightarrow MnSO_4 + 4H_2O + K^+$$

Add electrons to each half reaction so as to balance the charges.

$$Na_2C_2O_4 \rightarrow 2CO_2 + 2Na^+ + 2e^-$$
$$KMnO_4 + 8H^+ + SO_4^= + 5e^- \rightarrow MnSO_4 + 4H_2O + K^+$$

Multiply the first half reaction by 5 and the second half reaction by 2 to balance the electron change. Then add the two half reactions noting that the electrons on each side of the reaction cancel.

$$5Na_2C_2O_4 \rightarrow 10CO_2 + 10Na^+ + 10e^-$$
$$\underline{2KMnO_4 + 16H^+ + 2SO_4^= + 10e^- \rightarrow 2MnSO_4 + 8H_2O + 2K^+}$$
$$5Na_2C_2O_4 + 2KMnO_4 + 16H^+ + 2SO_4^= + \cancel{10e^-}$$
$$\rightarrow 10CO_2 + 10Na^+ + 2MnSO_4 + 8H_2O + 2K^+ + \cancel{10e^-}$$

Now balance the atoms not involved in the oxidation or reduction, combining them into the proper molecular form.

$$5Na_2C_2O_4 + 2KMnO_4 + 8H_2SO_4$$
$$\rightarrow 10CO_2 + 5Na_2SO_4 + 2MnSO_4 + 8H_2O + K_2SO_4$$

A quick check can be made by noting that eight $SO_4^=$ ions are on each side of the equation. A more complete check proves that the equation is properly balanced.

This method for balancing redox equations is quite useful for balancing an equation in the net ionic form. The following example illustrates this:

$$Cr_2O_7^= + Sn^{++} + H^+ \rightarrow Cr^{+++} + Sn^{+4} + H_2O$$

The two half reactions are

$$Cr_2O_7^= \rightarrow Cr^{+++}$$

and

$$Sn^{++} \rightarrow Sn^{+4}$$

Balance the atoms in each half reaction.

$$Cr_2O_7^= + 14H^+ \rightarrow 2Cr^{+++} + 7H_2O$$
$$Sn^{++} \rightarrow Sn^{+4}$$

Add electrons to balance the charge.

$$Cr_2O_7^= + 14H^+ + 6e^- \rightarrow 2Cr^{+++} + 7H_2O$$
$$Sn^{++} \rightarrow Sn^{+4} + 2e^-$$

Multiply the second half reaction by 3 to balance the electron change. Sum the two half reactions.

$$Cr_2O_7^= + 14H^+ + 6e^- \rightarrow 2Cr^{+++} + 7H_2O$$
$$3Sn^{++} \rightarrow 3Sn^{+4} + 6e^-$$

$$\overline{Cr_2O_7^= + 14H^+ + 3Sn^{++} + \cancel{6e^-} \rightarrow 2Cr^{+++} + 3Sn^{+4} + \cancel{6e^-} + 7H_2O}$$

The equation is now correctly balanced.

Either of the two methods for balancing redox equations will yield the correct answer. The student should adopt the method with which he feels most secure.

Oxidizing and Reducing Capacities

The capacity of a redox substance is measured by the weight of that substance which can be oxidized or reduced by an equivalent weight of another redox substance. The equivalent weight of a substance involved in a redox reaction is given as follows:

$$\text{equivalent weight} = \frac{\text{formula weight}}{n}$$

where n is the total number of electrons gained or lost per molecule. The equivalent weight of a substance therefore depends upon the reaction it undergoes. Some examples are shown in Table 9-1.

Table 9-1

Equivalent Weights of Some Redox Substances

Reaction	Equivalent weight of reactant
$Fe^{++} \rightarrow Fe^{+++} + e^-$	$\dfrac{Fe}{1}$
$Cr_2O_7^= + 14H^+ + 6e^- \rightarrow 2Cr^{+++} + 7H_2O$	$\dfrac{Cr_2O_7^=}{6}$
$MnO_4^- + 8H^+ + 5e^- \rightarrow Mn^{++} + 4H_2O$	$\dfrac{MnO_4^-}{5}$
$MnO_4^- + 4H^+ + 3e^- \rightarrow MnO_2 + 2H_2O$	$\dfrac{MnO_4^-}{3}$
$I_2 + 2e^- \rightarrow 2I^-$	$\dfrac{I_2}{2}$
$S_2O_3^= \rightarrow \frac{1}{2}S_4O_6^= + e^-$	$\dfrac{S_2O_3^=}{1}$

Redox Titrations

The theory of why redox reactions occur will be developed in the following sections of this chapter. The titration of an oxidizing agent with a reducing agent is used for the analysis of many substances. The end point of the titration can be determined either visually with a redox indicator or instrumentally with a potentiometer.

The titration of reducing agents with potassium permanganate in acid solutions requires no added indicator for a visual end point detection. The intense color of the permanganate ion serves as its own indicator of an excess amount of oxidant. Less than one drop of excess $0.10N$ $KMnO_4$ can be detected in 100 ml of titration solution provided the solution is colorless or nearly colorless prior to the equivalence point.

Calculations of Redox Titrations

The equation used to calculate the results of a redox titration is the same as Equation (6-11).

$$\%Y = \frac{V_t \times N_t \times \dfrac{\text{formula weight of } Y}{n \times 1000} \times 100}{SW}$$

where V_t and N_t are the volume in ml of the titrant and normality of the titrant respectively, n is the number of electrons gained or lost per formula weight of Y and SW is the sample weight.

Reasonable care must be taken in stating the normality of the titrant. The normality of a substance pertains to a specific half reaction. Some species undergo more than a single half reaction depending on the reaction conditions. A good example of this is potassium permanganate. In a strong acid solution the permanganate ion undergoes a five electron change:

$$MnO_4^- + 8H^+ + 5e^- \rightarrow Mn^{++} + 4H_2O$$

The equivalent weight of potassium permanganate in an acid solution is one-fifth its molecular weight or 31.61 g. In a neutral solution the permanganate ion undergoes a three electron change:

$$MnO_4^- + 4H^+ + 3e^- \rightarrow MnO_2 + 2H_2O$$

In a neutral solution the equivalent weight of potassium permanganate is one-third its molecular weight or 52.68 g. In a strongly basic solution potassium permanganate only undergoes a one electron change:

$$MnO_4^- + e^- \rightarrow MnO_4^=$$

In this medium the equivalent weight is 158.04 g. Since the equivalent weight of potassium permanganate changes depending upon the reaction, the normality of a standard solution of it will also change. Example 9-1 will illustrate the type problem encountered.

EXAMPLE 9-1: A potassium permanganate solution was standardized against primary standard sodium oxalate in sulfuric acid solution; 0.2820 g requiring 42.15 ml of the permanganate. The standardized solution was used to titrate a 0.1423 g sample of impure manganese chloride dissolved in a neutral solution. This titration required 37.15 ml of permanganate. Calculate the percent purity of the manganese chloride. The reactions involved are the following:

standardization: $5Na_2C_2O_4 + 2KMnO_4 + 3H_2SO_4$

$$\rightarrow K_2SO_4 + 10CO_2 + 2MnSO_4 + 8H_2O$$

titration: $3MnCl_2 + 2ZnO + 2KMnO_4$

$$\rightarrow 5MnO_2 + 2KCl + 2ZnCl_2$$

The normality of the permanganate in acid solution is calculated by the equation

$$N = \frac{0.2820 \text{ g}}{\dfrac{Na_2C_2O_4}{2000} \times 42.15 \text{ ml}}$$

$$= 0.0999N \text{ (acid solution)}$$

However, the normality of the solution in the titration of the manganese chloride is only three-fifths as great.

$$N = 0.0999 \times \frac{3}{5}$$

$$= 0.0599N \text{ (neutral solution)}$$

The percent manganese chloride in the sample is given as follows:

$$\%MnCl_2 = \frac{37.15 \text{ ml} \times 0.0599N \times \dfrac{MnCl_2}{2000} \times 100}{0.1423}$$

$$= 98.4\%$$

Electrode Potentials

If a noble metal such as platinum is placed in a solution containing the oxidized and reduced forms of an ion, the noble metal acquires an electric charge which varies with the nature of the element and the concentration of ions in the redox system. The platinum metal is known as an electrode. The charge on the platinum is a measure of the *electrode potential* of the system. If the platinum electrode is in contact with a solution of ferric and ferrous ions at equilibrium (as shown below) a certain potential develops.

$$Fe^{+++} + e^- \rightleftharpoons Fe^{++}$$

If ferric ions are added, the equilibrium is shifted to the right and the electrode potential becomes more positive. If ferrous ions are added, the equilibrium is shifted to the left and the electrode potential becomes more negative. The electrode potential is, therefore, a measure of the redox ability of the system. The noble metal electrode is called an indicating electrode.

Similarly, if a conducting metal such as a bar of zinc is dipped into a solution of its own ions, an electric charge develops. The zinc bar is also known as an electrode. Again a reversible reaction occurs at the surface of the zinc electrode which comes to equilibrium.

$$Zn^{++} + 2e^- \rightleftharpoons Zn$$

An increase in zinc ions will shift the equilibrium to the right and raise the electrode potential.

Unfortunately, it is impossible to measure the potential of a single electrode. However, the potential difference between two electrodes can be easily measured. This potential difference is called the cell potential. If many cell potentials are measured, it is possible to assign single electrode potentials (half cell potentials) compared to a standard reference electrode. In this manner the relative potential of various half cells can be determined.

The standard hydrogen electrode (SHE) or normal hydrogen electrode (NHE) is used as the reference electrode. The SHE is a thin sheet of platinum coated with a thin layer of finely divided platinum metal known as platinum black. The platinum black absorbs hydrogen gas until it is saturated and then hydrogen is bubbled onto the electrode at 1 atmosphere pressure while the electrode is in contact with a solution containing hydrogen ions at unit activity. The potential of this

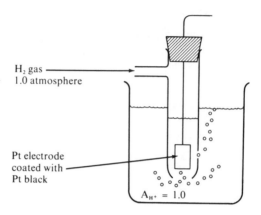

Figure 9-1

A standard hydrogen electrode.

electrode is given the arbitrary value of 0.000 volts. Figure 9-1 illustrates a typical hydrogen electrode.

The standard hydrogen electrode is rarely used in electrode potential measurements due to the difficulty of preparation and of maintaining conditions of equilibrium (H_2 pressure of 1.0 atmosphere, unit activity of hydrogen ions). Instead a calomel electrode or a silver-silver chloride electrode is usually used as the reference. The saturated calomel electrode (SCE) contains elemental mercury and a paste made of mercurous chloride and metallic mercury. The paste is in contact with a saturated solution of potassium chloride. The SCE has a potential of +0.2458 volts vs. SHE at 25°C. The silver-silver chloride electrode consists of a silver electrode immersed in a saturated potassium chloride solution which is also saturated with silver chloride. This electrode has a potential of +0.222 volts vs. SHE at 25°C.

Standard Electrode Potentials

Not all chemists agree upon the convention for writing the direction of an electrode reaction and the subsequent sign of its potential. The electrode potential convention used in this text is the Gibbs-Stockholm convention adopted at the seventeenth conference of the International Union of Pure and Applied Chemistry in Stockholm, in 1953. The rules for this convention are the following:

1. Write the redox half reactions as reductions.

$$Fe^{+++} + e^- \rightleftharpoons Fe^{++}$$

$$Cu^{++} + 2e^- \rightleftharpoons Cu$$

$$Zn^{++} + 2e^- \rightleftharpoons Zn$$

$$2H^+ + 2e^- \rightleftharpoons H_2$$

2. Write the sign of the potential as plus if the oxidized form of the couple is a better oxidizing agent than hydrogen ion. Write the sign as minus if hydrogen ion is a better oxidizing agent than the oxidized form of the couple being compared.
3. The standard electrode potential $E°$ is the potential of the electrode when each substance involved in the half reaction is at unit activity. When the reacting species are dissolved in a solvent, the activity is assumed to be approximately equal to the concentration in moles per liter. If the reactant is a gas, its activity is equal to the partial pressure of the gas in atmospheres. If the reactant exists in a second phase as a pure solid or liquid, its activity is assumed to be unity because the concentration of substances in their pure phase is constant and independent of the quantity of the second phase. If the solvent water is involved in the half cell process, its concentration remains essentially unchanged and its activity is included in the standard electrode potential $E°$. $E°$ values are expressed compared to SHE. The standard electrode potential in a half reaction such as

$$MnO_4^- + 8H^+ + 5e^- \rightleftharpoons Mn^{++} + 4H_2O$$

is the potential of the electrode when not only the activities of MnO_4^- and Mn^{++} are unity but when the activity of H^+ is also unity.

Some standard electrode potentials are listed in Table 9-2. A more complete listing is found in Appendix B. The higher the value of $E°$, the greater the oxidizing power of the system. Therefore, the strongest oxidizing agents are the oxidized form of the couples at the top of the table. The strongest reducing agents are the reduced form of the couples at the bottom of the table. In general the oxidized form of a couple should oxidize the reduced form of any couple below it in the

Table 9-2

Some Selected Standard Electrode Potentials

Half-reaction	$E°$, Volts
$F_2 + 2H^+ + 2e = 2HF$	3.06
$S_2O_8^= + 2e = 2SO_4^=$	2.01
$Ce^{+4} + e = Ce^{+3}$	1.61
$MnO_4^- + 8H^+ + 5e = Mn^{++} + 4H_2O$	1.51
$Cr_2O_7^= + 14H^+ + 6e = 2Cr^{+++} + 7H_2O$	1.33
$MnO_2 + 4H^+ + 2e = Mn^{++} + 2H_2O$	1.23
$O_2 + 4H^+ + 4e = 2H_2O$	1.229
$IO_3^- + 6H^+ + 5e = \frac{1}{2}I_2 + 3H_2O$	1.195
$VO_2^+ + 2H^+ + e = VO^{++} + H_2O$	1.00
$2Hg^{++} + 2e = Hg_2^{++}$	0.920
$Ag^+ + e = Ag$	0.800
$Fe^{+++} + e = Fe^{++}$	0.771
$H_3AsO_4 + 2H^+ + 2e = H_3AsO_3 + H_2O$	0.559
$I_2 + 2e = 2I^-$	0.535
$Cu^{++} + 2e = Cu$	0.337
$Hg_2Cl_2 + 2e = 2Hg + 2Cl^-$	0.268
$AgCl + e = Ag + Cl^-$	0.222
$Sn^{+4} + 2e = Sn^{++}$	0.15
$2H^+ + 2e = H_2$	0.000
$Pb^{++} + 2e = Pb$	−0.126
$Sn^{++} + 2e = Sn$	−0.136
$Ni^{++} + 2e = Ni$	−0.250
$Co^{++} + 2e = Co$	−0.277
$Cd^{++} + 2e = Cd$	−0.403
$Zn^{++} + 2e = Zn$	−0.763
$Li^+ + e = Li$	−3.04

table. However, a catalyst may be necessary to bring about the indicated redox reaction in a reasonable time.

The Nernst Equation

The standard potential of an electrode is the potential difference between the SHE and the electrode when both the oxidized and reduced forms of the couple are at unit activity. The potential of the electrode changes with changes in concentration of the two forms. The potential of the electrode can be determined experimentally by measuring the potential difference between it and a reference electrode having a known potential. A relationship between the electrode potential and the activity of the ions in equilibrium with the electrode was derived by W. Nernst in 1889 and is known as the Nernst equation.

For the general half reaction where a molecules of the oxidized form are in equilibrium with b molecules of the reduced form,

$$a \text{ (oxidized form)} + ne^- \rightleftharpoons b \text{ (reduced form)}$$

The Nernst equation is written as

$$E = E° + \frac{RT}{nF} \ln \frac{a \text{ oxid}^a}{a \text{ red}^b}$$

Where: E is the potential of the electrode in volts.
         $E°$ is the standard electrode potential.
         R is the gas constant, 8.314 joules per degree-mole.
         T is the absolute temperature.
         n is the number of electrons involved.
         F is the Faraday, 96,493 coulombs.
         $a \text{ oxid}^a$ is the activity of the oxidized form raised to the power equal to its coefficient in the balanced equation.
         $a \text{ red}^b$ is the activity of the reduced form raised to the power equal to its coefficient in the balanced equation.

The Nernst equation may also be written as

$$E = E° + \frac{RT}{nF} \ln \frac{1}{K_{eq}}$$

where K_{eq} is the equilibrium constant for the half reaction written as a reduction and all other terms are as defined previously. In this form, it should be clear that the activity of all species involved in the half reaction must be equal to unity for the electrode potential to be defined as the standard electrode potential.

Changing from natural logarithms to common logarithms and assuming that activities can be expressed as concentrations, the Nernst equation at 25°C becomes

$$E = E° + \frac{0.059}{n} \log \frac{[oxid]^a}{[red]^b}$$

The use of the Nernst equation is illustrated by the following examples.

EXAMPLE 9-2:   Calculate the potential of a platinum electrode immersed in a solution containing $0.01 M$ Fe^{+++} and $0.10 M$ Fe^{++}.

$$E = E° + \frac{0.059}{1} \log \frac{[Fe^{+++}]}{[Fe^{++}]}$$

$$= 0.771 + 0.059 \log \frac{10^{-2}}{10^{-1}}$$

$$= 0.771 + (0.059)(-1)$$

$$= 0.712 \text{ volts vs. SHE}$$

EXAMPLE 9-3:   Calculate the potential of a zinc electrode immersed in a solution containing 4.2×10^{-2} M Zn^{++}.

$$E = E° + \frac{0.059}{2} \log \frac{[Zn^{++}]}{[Zn]}$$

$$= -0.763 + \frac{0.059}{2} \log 4.2 \times 10^{-2}$$

$$= -0.763 + \frac{(0.059)}{2} (-1.38)$$

$$= -0.763 - 0.041$$

$$= 0.804 \text{ volts vs. SHE}$$

EXAMPLE 9-4:   Calculate the solubility product constant for silver oxalate, $Ag_2C_2O_4$, from the following data: A silver electrode immersed in a saturated aqueous solution of silver oxalate and a saturated calomel electrode (E = +0.246 volts vs. SHE) have a potential difference of 0.343 volts with the silver electrode being more positive than the calomel electrode.

To calculate the silver ion concentration it is necessary to compute the silver electrode potential vs SHE.   The stick diagram in Figure 9-2 will aid in computing that potential.

Since the silver electrode potential is 0.343 volts more positive than the SCE, its potential is +0.589 volts vs. SHE.   This potential

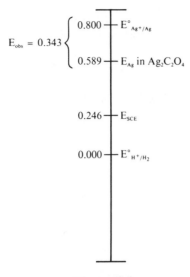

Figure 9-2

is used with the Nernst equation to calculate the silver ion concentration.

$$0.589 = 0.800 + 0.059 \log [Ag^+]$$

$$\log [Ag^+] = \frac{-0.211}{0.059} = -3.57$$

$$[Ag^+] = 2.7 \times 10^{-4}$$

The oxalate ion concentration in a saturated solution of silver oxalate is one-half the silver ion concentration.

$$[C_aO_4^=] = 1.35 \times 10^{-4}$$

$$K_{sp} = [Ag^+]^2 [C_2O_4^=]$$

$$= (2.7 \times 10^{-4})^2 (1.35 \times 10^{-4})$$

$$= 9.8 \times 10^{-12}$$

The potential of an electrode for a couple in which the hydrogen ion is involved depends upon the pH of the solution. Example 9-5 illustrates the effect.

EXAMPLE 9-5: Calculate the potential of a platinum electrode in a solution containing $0.1\,M$ $Cr_2O_7^=$ and $0.1\,M$ Cr^{+++} at (a) $[H^+] = 1\,M$, (b) $[H^+] = 2M$, and (c) $[H^+] = 1 \times 10^{-3}M$.

$$Cr_2O_7^= + 14H^+ + 6e^- \rightleftharpoons 2Cr^{+++} + 7H_2O \quad E° = +1.33\text{ v}$$

(a) $\quad E = 1.33 + \dfrac{0.059}{6} \log \dfrac{[Cr_2O_7^=][H^+]^{14}}{[Cr^{+++}]^2}$

$\quad\quad = 1.33 + \dfrac{0.059}{6} \log \dfrac{[Cr_2O_7^=]}{[Cr^{+++}]^2} + \dfrac{(14)(.059)}{6} \log [H^+]$

$\quad\quad = 1.33 + \dfrac{0.059}{6} \log \dfrac{0.1}{(0.1)^2} + \dfrac{(14)(0.059)}{6} \log 1$

$\quad\quad = 1.33 + .01 + 0$

$\quad\quad = 1.34\text{ volts vs. SHE}$

(b) $\quad E = 1.33 + .01 + \dfrac{(14)(0.059)}{6} \log 2$

$\quad\quad = 1.33 + .01 + .04$

$\quad\quad = 1.38\text{ volts vs. SHE}$

(c) $\quad E = 1.33 + .01 + \dfrac{(14)(0.059)}{6} \log 1 \times 10^{-3}$

$\quad\quad = 1.33 + .01 - .41$

$\quad\quad = 0.93\text{ volts vs. SHE}$

Cell Potentials

A redox reaction is a combination of two half reactions. Since standard electrode potentials are a measure of the oxidizing strength of a couple, the difference between two standard electrode potentials measures the driving force for the redox reaction between the two couples. A cell representing a spontaneous electrochemical process is known as a galvanic cell.

Consider two beakers, one containing a platinum electrode immersed in a solution of ferric and ferrous ions both at $1M$ and the second beaker containing a zinc electrode immersed in a solution containing $1M$ zinc ions. The standard electrode potential for the ferric-ferrous couple is more positive than the standard electrode potential for zinc ion-zinc couple. Therefore, ferric ion should oxidize zinc, the reaction being

$$2Fe^{+++} + Zn \rightarrow 2Fe^{++} + Zn^{++}$$

If the two electrodes are connected with a wire as shown in Figure 9-3 reactions will begin to occur at the surface of each electrode.

At the zinc electrode the equilibrium

$$Zn \rightarrow Zn^{++} + 2e^-$$

is established. A small amount of the zinc electrode dissolves to form zinc ions. The electrons formed in the process flow through the wire to the platinum electrode where the charge is dissipated by the reaction

$$Fe^{+++} + e^- \rightarrow Fe^{++}$$

This exchange of electrons can occur only to an infinitesimal extent because an immediate charge imbalance is created. An excess of zinc ions occurs at the interface of the zinc electrode which resists the formation of additional positive ions. Likewise at the platinum electrode an excess of negative ions are formed as the result of the $+3$ charged ferric ion being converted into the $+2$ charged ferrous ion. A steady-state situation is rapidly established with no measurable change in composition.

If a salt bridge (see Figure 9-4) is introduced into the circuit, current

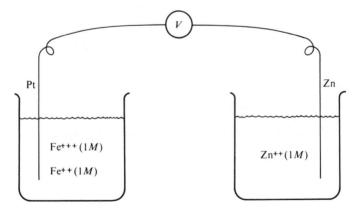

Figure 9-3

A nonfunctioning galvanic cell.

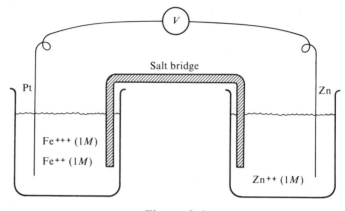

Figure 9-4

A galvanic cell.

can continue to flow and a cell potential or voltage can be measured. A simple salt bridge consists of an inverted U-tube filled with a mixture of a salt solution and agar which has been formed into a gel. The salt bridge allows a transfer of ions by migration between the solutions to occur without allowing the two solutions to mix. This maintains an electrical balance and a current will now flow through the external circuit. The potential of a galvanic cell changes as current flows because as the concentrations of the species making up the half reactions change the electrode potentials change. The initial potential of the galvanic cell in Figure 9-4 is equal to the difference between the two respective standard electrode potentials. The initial potential of this cell is 1.534 volts and is known as the *standard potential* of the cell. This potential can be measured with a voltmeter. If the circuit remains closed, the redox reaction will proceed to equilibrium with essentially all the ferric ions being converted into ferrous ions.

The following examples are given.

EXAMPLE 9-6: Calculate the standard cell potential for the reaction

$$2Fe^{+++} + Sn^{++} \rightarrow 2Fe^{++} + Sn^{+4}$$

$$2Fe^{+++} + 2e^- = 2Fe^{++} \qquad E° = 0.77 \text{ v}$$

$$\underline{-(Sn^{+4} + 2e^- = Sn^{+2}) \qquad -(E° = 0.15 \text{ v})}$$

$$2Fe^{+++} + Sn^{++} \rightarrow 2Fe^{++} + Sn^{+4} \qquad E° = 0.62 \text{ v}$$

The standard potential for the reaction is 0.62 volts.

EXAMPLE 9-7: Calculate the potential of a cell composed of a copper electrode immersed in $0.04M$ $CuCl_2$ and a zinc electrode immersed in $2.0M$ $ZnCl_2$.

To determine the cell potential when the concentrations are not $1M$, the potential of each electrode must be calculated from the Nernst equation. The difference between the two electrode potentials is the cell potential.

$$E_{Cu^{++}/Cu} = 0.337 + \frac{0.059}{2} \log 0.04$$

$$= 0.337 - 0.041$$

$$= 0.296 \text{ v}$$

$$E_{Zn^{++}/Zn} = -0.763 + \frac{0.059}{2} \log 2.0$$

$$= -0.763 + .009$$

$$= -0.754 \text{ v}$$

$$E_{cell} = E_{Cu^{++}/Cu} - E_{Zn^{++}/Zn} = 0.296 - (-0.754)$$

$$= 1.050 \text{ v}$$

Equilibrium Constants from Cell Potentials

As indicated in the previous section, when current is allowed to flow through a galvanic cell, the redox reaction will proceed to equilibrium. The potentials of the two electrodes will change as the reaction progresses. As the current flows the electrode potentials will become closer together. When equilibrium is reached, no more current flows and the potentials of both electrodes are equal.

Consider the reaction

$$2Fe^{+++} + Sn^{++} \rightarrow 2Fe^{++} + Sn^{+4}$$

At equilibrium the potentials of both half cells are equal. The potential of the iron couple is given by

$$E_{Fe} = E_{Fe}^{\circ} + \frac{0.059}{1} \log \frac{[Fe^{+++}]}{[Fe^{++}]}$$

and the potential of the tin couple is given by

$$E_{Sn} = E_{Sn}^\circ + \frac{0.059}{2} \log \frac{[Sn^{+4}]}{[Sn^{++}]}$$

Since the two potentials are equal, we see that

$$E_{Fe}^\circ + \frac{0.059}{1} \log \frac{[Fe^{+++}]}{[Fe^{++}]} = E_{Sn}^\circ + \frac{0.059}{2} \log \frac{[Sn^{+4}]}{[Sn^{++}]}$$

Therefore,

$$E_{Fe}^\circ - E_{Sn}^\circ = \frac{0.059}{2} \log \frac{[Sn^{+4}]}{[Sn^{++}]} - \frac{0.059}{1} \log \frac{[Fe^{+++}]}{[Fe^{++}]}$$

$$= \frac{0.059}{2} \log \frac{[Sn^{+4}]}{[Sn^{++}]} - \frac{0.059}{2} \log \frac{[Fe^{+++}]^2}{[Fe^{++}]^2}$$

Combining the two log terms, we have

$$E_{Fe}^\circ - E_{Sn}^\circ = \frac{0.059}{2} \log \frac{[Sn^{+4}][Fe^{++}]^2}{[Sn^{++}][Fe^{+++}]^2}$$

The log term is the equilibrium constant expression. Therefore,

$$E_{Fe}^\circ - E_{Sn}^\circ = \frac{0.059}{2} \log K_{eq}$$

$$\log K_{eq} = \frac{0.771 - 0.15}{\dfrac{0.059}{2}} = 21$$

$$K_{eq} = 10^{21}$$

The large value for the equilibrium constant indicates that the reaction goes far towards completion. This reaction is utilized in the reduction of iron to the plus two oxidation state prior to titration with dichromate in the volumetric determination of iron (Experiment 11).

The general equation relating the equilibrium constant to the standard electrode potentials is

$$\log K_{eq} = \frac{n(E_1^\circ - E_2^\circ)}{0.059}$$

where n is the number of electrons transferred in the reaction.

Potentiometric Titrations

In a potentiometric titration the potential of an electrode is plotted as a function of the volume of titrant added. Potentiometric titrations are applicable to acid-base, complexometric, precipitation, and redox reactions. Advantages of potentiometric titrations include: (1) They can be used for titration of brightly colored species where visual methods may not be applicable. (2) They eliminate the need for an indicator. (3) They eliminate the error of matching color changes at the equivalence point. (4) For many reactions the equivalence point can be determined very accurately. (5) The equipment is relatively simple.

A potentiometric titration is normally performed in an open beaker. An indicating electrode often made of platinum and a reference electrode, which is usually a saturated calomel or a silver-silver chloride electrode, are immersed in the solution and are connected to a potentiometer. A potentiometer is a device designed to measure the potential difference between two electrodes without drawing any appreciable current. Commercial pH meters are designed to be used as potentiometers by replacing the hydrogen ion sensitive glass electrode with an indicating electrode. The solution should be stirred continuously during the course of the titration and this is most conveniently accomplished with a magnetic stirrer and a magnetic stirring bar. This apparatus is shown in Figure 9-5.

Consider the titration of 50.0 ml of $0.1000N$ iron(II) with $0.1000N$ cerium(IV). The half reactions are the following:

$$Ce^{+4} + e^- = Ce^{+3} \qquad E° = 1.61 \text{ v}$$

$$Fe^{+3} + e^- = Fe^{+2} \qquad E° = 0.771 \text{ v}$$

At the beginning of the titration the solution contains only ferrous iron. Although the indicating electrode does measure a potential, this potential is not defined and cannot be calculated from the Nernst equation. After the addition of ceric solution but prior to the equivalence point there exists a mixture of ferrous, ferric, and cerous ions. The potential of the electrode is calculated from the Nernst equation for the iron couple.

At 10.0 ml titration:

$$[Fe^{+++}] = \frac{10.0 \text{ ml} \times 0.1000N}{60 \text{ ml}}$$

$$[Fe^{++}] = \frac{40.0 \text{ ml} \times 0.1000N}{60 \text{ ml}}$$

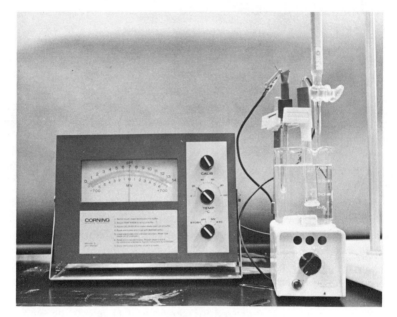

Figure 9-5

Apparatus for a potentiometric titration.

$$E = E° + \frac{0.059}{1} \log\frac{[Fe^{+++}]}{[Fe^{++}]}$$

$$= 0.771 + 0.059 \log\frac{10}{40}$$

$$= 0.771 - 0.036$$

$$= 0.735 \text{ v vs. SHE or } 0.489 \text{ v vs. SCE}$$

At 25.0 ml titration:

$$[Fe^{+++}] = [Fe^{++}]$$

$$E = E°$$

$$= 0.771 \text{ v vs. SHE or } 0.525 \text{ v vs. SCE}$$

At 40.0 ml titration:

$$E = 0.771 + 0.059 \log\frac{40}{10}$$

$$= 0.771 + .036$$

$$= 0.807 \text{ v vs. SHE or } 0.561 \text{ v vs. SCE}$$

At 50.0 ml titration, which is the equivalence point, the solution contains primarily cerous and ferric ions. However, an equilibrium concentration of ceric and ferrous ions also exists. Therefore, either redox couple can be used to express the potential of the indicating electrode.

$$E_{Fe} = 0.771 + \frac{0.059}{1} \log \frac{[Fe^{+++}]}{[Fe^{++}]}$$

$$E_{Ce} = 1.61 + \frac{0.059}{1} \log \frac{[Ce^{+4}]}{[Ce^{+++}]}$$

Since both equations express the potential of the indicating electrode, the two potentials must be equal to the end point potential E. The sum of the two equations gives

$$2E = 0.771 + 1.61 + \frac{0.059}{1} \log \frac{[Fe^{+++}][Ce^{+4}]}{[Ce^{+++}][Fe^{++}]}$$

At the equivalence point $[Ce^{+++}] = [Fe^{+++}]$ because they are formed by the reaction on a one to one mole basis. At the exact equivalence point, $[Ce^{+4}] = [Fe^{++}]$ because the total moles of ceric added at the equivalence point by definition equal the total moles of ferrous originally present, and the moles which have reacted are equal leaving the moles unreacted also equal. Consequently the log term in the above equation reduces to 1.

$$2E = 0.771 + 1.61 + \frac{0.059}{1} \log 1$$

$$= 0.771 + 1.61 + 0$$

$$E = \frac{0.771 + 1.61}{2} = 1.19 \text{ v vs. SHE or } 0.94 \text{ v vs. SCE}$$

After the equivalence point is passed, the solution contains predominately ceric, cerous, and ferric ions. A small equilibrium concentration of ferrous ion is present but its concentration is not directly known. The ceric and cerous concentrations are readily calculated with the ceric concentration being equal to the excess titrant added and the cerous concentration equal to the original ferrous present (both con-

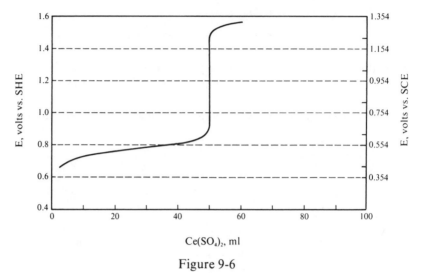

Figure 9-6

Titration curve for $0.100M$ FeCl$_2$ with $0.100M$
Ce(SO$_4$)$_2$ in $1M$ HNO$_3$.

centrations being corrected for the dilution effect). The potential is, therefore, calculated by the Nernst equation for the cerium couple.

At 60.0 ml titration:

$$E = E° + \frac{0.059}{1} \log\frac{[Ce^{+4}]}{[Ce^{+++}]}$$

$$= 1.61 + 0.059 \log\frac{10}{50}$$

$$= 1.61 - 0.04_1$$

$$= 1.57 \text{ v vs. SHE or } 1.32 \text{ v vs. SCE}$$

The titration curve for this reaction is shown in Figure 9-6. Observed potentials vs. saturated calomel electrode are shown on the right and standard potentials on the left.

Equivalence Point Potentials

The equation derived in the previous section for the potential at the equivalence point of a redox titration can be generalized as follows:

$$E = \frac{n_1 E_1^\circ + n_2 E_2^\circ}{n_1 + n_2}$$

where n_1 is the number of electrons involved in the half reaction represented by E_1° and n_2 is the number of electrons involved in the half reaction represented by E_2°. This equation is valid for all reactions in which the hydrogen ion is not involved and where each couple has a molar ratio of one between the species in the oxidized form and the reduced form. The following two examples illustrate the exceptions.

EXAMPLE 9-8: Calculate the equivalence point potential for the titration of ferrous ion with permanganate ion in an acid solution.

$$E = E_{Fe}^\circ + \frac{0.059}{1} \log \frac{[Fe^{+++}]}{[Fe^{++}]}$$

and

$$E = E_{MnO_4^-}^\circ + \frac{0.059}{5} \log \frac{[MnO_4^-][H^+]^8}{[Mn^{++}]}$$

In order to combine the logarithmic terms, the second equation must be multiplied through by 5 giving

$$5E = 5E_{MnO_4^-}^\circ + \frac{0.059}{1} \log \frac{[MnO_4^-][H^+]^8}{[Mn^{+++}]}$$

Adding this to the equation for E above gives

$$6E = E_{Fe}^\circ + 5E_{MnO_4^-}^\circ + 0.059 \log \frac{[MnO_4^-][H^+]^8[Fe^{+++}]}{[Mn^{++}][Fe^{++}]}$$

The stoichiometry of the reaction requires that at the equivalence point

$$[Fe^{+++}] = 5[Mn^{++}]$$

and

$$[Fe^{++}] = 5[MnO_4^-]$$

Substitution of these yields

$$6E = E_{Fe}^{\circ} + E_{MnO_4^-}^{\circ} + 0.059 \log \frac{[MnO_4^-][H^+]^8 5[Mn^{++}]}{[Mn^{++}]5[MnO_4^-]}$$

$$E = \frac{E_{Fe}^{\circ} + 5E_{MnO_4^-}^{\circ}}{6} + \frac{0.059}{6} \log[H^+]^8$$

$$= 1.39 + \frac{(8)(0.059)}{6} \log[H^+]$$

EXAMPLE 9-9: Calculate the equivalence point potential for the titration of ferrous ion with dichromate ion in an acid solution.

$$E = E_{Fe}^{\circ} + 0.059 \log \frac{[Fe^{+++}]}{[Fe^{++}]}$$

$$E = E_{Cr_2O_7^-}^{\circ} + \frac{0.059}{6} \log \frac{[Cr_2O_7^-][H^+]^{14}}{[Cr^{+++}]^2}$$

Proceeding as in Example 9-8, we have

$$7E = E_{Fe}^{\circ} + 6E_{Cr_2O_7^-}^{\circ} + 0.059 \log \frac{[Fe^{+++}][Cr_2O_7^-][H^+]^{14}}{[Fe^{++}][Cr^{+++}]^2}$$

At the equivalence point

$$[Fe^{+++}] = 3[Cr^{+++}]$$

$$[Fe^{++}] = 6[Cr_2O_7^-]$$

Substituting these values yields

$$E = \frac{E_{Fe}^{\circ} + 6E_{Cr_2O_7^-}^{\circ}}{7} + \frac{(14)(0.059)}{7} \log[H^+] - \frac{0.059}{7} \log 2[Cr^{+++}]$$

In Example 9-8, we see that the equivalence point potential depends upon the pH of the solution. In Example 9-9, the equivalence point depends not only upon the pH of the solution but also upon the chromic ion concentration. The dependence upon the chromic ion concentration is due to the fact that the number of moles of chromium in the oxidized and reduced forms of the couple are not equal.

Selection of Visual Indicators for Redox Titrations

As mentioned previously, titrations with potassium permanganate solutions usually do not require the use of an indicator. The permanganate ion serves as its own indicator.

Starch indicator is used in redox titrations involving iodine. An aqueous starch solution forms a deep blue complex in the presence of iodine and iodide ion. For titrations with iodine, starch solution is added at the start of the titration. Iodide ion is produced as the reduction product. The blue complex appears when the first drop of excess iodine remains in solution. For titration of iodine solutions with sodium thiosulfate solution, the titrant is added until the iodine reaches a pale straw-like color. Starch indicator is added and the titration is then continued until the blue color disappears.

The above two end point indicators do not depend on the change in potential of the solution, although the completeness of the reaction and the sharpness of the end point do. Most redox titration end points are detected with a redox indicator. A redox indicator is an organic dye which can be oxidized or reduced by the titrating solution. The colors of the dye in the oxidized and reduced states are different. The half reaction and Nernst equation for the indicator are

$$\text{Oxid}_{\text{Ind}} + ne^- \rightleftharpoons \text{Red}_{\text{Ind}}$$

$$E_{\text{Ind}} = E_{\text{Ind}}^{\circ} + \frac{0.059}{n} \log \frac{[\text{Oxid}_{\text{Ind}}]}{[\text{Red}_{\text{Ind}}]}$$

The potential of the solution will determine E_{Ind}. Therefore, the ratio of oxidized to reduced forms is also determined by the potential of the solution. As the ratio of the two colored forms of the indicator change, the visible color changes. Assuming the eye can see 1 part of a second color in the presence of 10 parts of the first, the visible color change will occur when the ratio changes from 10/1 to 1/10. If the indicator is originally in the reduced form the first detectable change in color will occur at

$$E = E_{\text{Ind}}^{\circ} + \frac{0.059}{n} \log \frac{1}{10}$$

$$= E_{\text{Ind}}^{\circ} - \frac{0.059}{n}$$

The last detectable change occurs at

$$E = E^{\circ}_{Ind} + \frac{0.059}{n} \log \frac{10}{1}$$

$$= E^{\circ}_{Ind} + \frac{0.059}{n}$$

Thus, a change of $[(2 \times 0.059)/n]$ volts is required for the total visible color change to occur. This is the transition range for the indicator. For n equal to one, 0.118 volts is the transition range. To be used in a redox titration, the transition range for the indicator must fall within the steep equivalence point break for the redox reaction. In addition, the indicator redox reaction must be rapid and reversible. A minimal concentration of indicator should be employed for a sharp end point because the titrant must react with the indicator. For accurate determinations an indicator blank should be subtracted to account for the volume of titrant required to react with the indicator. An alternative to an indicator blank correction is to standardize the titrant using the same method and the same amount of indicator. Table 9-3 lists some commonly employed redox indicators.

Table 9-3

Some Redox Indicators

| | Color | | | |
Indicator	Reduced form	Oxidized form	Solution conditions	E°, volts
Tris(5–nitro–1,10 phenanthroline) iron(II) sulfate (nitro ferroin)	red	pale blue	$1 M\ H_2SO_4$	+1.25
1,10 phenanthroline iron(II) sulfate (ferroin)	red	pale blue	$1 M\ H_2SO_4$	+1.06
diphenylamine sulfonic acid	colorless	purple	dilute acid	+0.84
diphenylamine	colorless	violet	dilute acid	+0.76
methylene blue	blue	colorless	$1 M$ acid	+0.53
indigo tetrasulfonate	colorless	blue	$1 M$ acid	+0.36
phenosafranine	colorless	red	$1 M$ acid	+0.28

Measurement of pH

The pH of a solution can be measured by a potentiometer or a pH meter provided the potential of the indicating electrode is a function of the pH of the solution and the potential of the reference electrode is independent of the pH.

A hydrogen electrode could serve as an indicating electrode but is too bulky and inconvenient to be practical. A *glass electrode* is nearly always used as the measuring electrode for pH measurements.

Figure 9-7 shows a diagram of a typical glass electrode. It consists of a silver wire coated with silver chloride immersed in a dilute solution of hydrochloric acid surrounded by a bulb containing a thin membrane of a special glass. The silver-silver chloride internal reference electrode has a constant potential, but a potential difference develops across the glass membrane. This potential is a function of the difference in hydrogen ion activity of the hydrochloric acid concentration inside the electrode and the hydrogen ion activity in the test solution. The potential results from the migration of hydrogen ions from the solution into the outer layer of the glass membrane as represented by the equation

$$-O-Si-O^{\ominus}-Na^{\oplus} + H^+ \rightleftharpoons -O-Si-O^{\ominus}-H^{\oplus} + Na^+$$

The more hydrogen ions present in the solution, the more migration

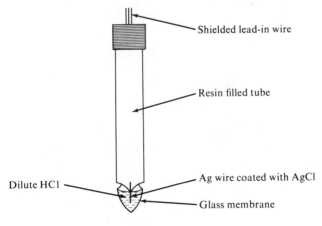

Figure 9-7

A typical glass electrode.

occurs. This results in a change in the membrane potential. A similar reaction occurs on the inside of the membrane. However, since the hydrogen ion concentration is constant on the inside, a constant inner membrane potential exists. Therefore, the potential difference *across* the membrane will depend only upon the hydrogen ion concentration (activity) in the test solution.

The potential of a glass electrode at 25°C is expressed by the equation

$$E = K + 0.059 \text{ pH}$$

where K is a constant containing among other components an *asymmetry potential* which varies from day to day. The asymmetry potential is that small potential which develops across the glass membrane when identical solutions are on each side of the membrane. It is found to depend on the composition of the glass membrane, its shape, and the way it was made. It is believed in part to be due to strains in the glass membrane. Because the asymmetry potential varies, even for a particular electrode, the measuring system (pH meter) must be calibrated at the time of each measurement. To calibrate the pH meter immerse both the glass electrode and the reference electrode in a buffer solution of a known pH and adjust the pH meter until it reads the appropriate value. Commercial pH meters are equipped with a calibration control to perform the necessary adjustment.

A saturated calomel electrode (Figure 9-8) or a silver-silver chloride

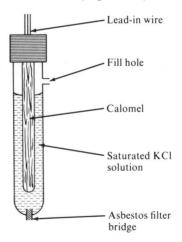

Figure 9-8

A saturated calomel electrode.

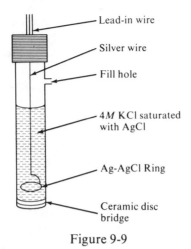

Figure 9-9

A silver-silver chloride reference electrode.

electrode (Figure 9-9) is used as the reference electrode. A calomel electrode contains a tube containing mercury metal and mercurous chloride paste (calomel). This is in contact with a saturated solution of potassium chloride which serves as a bridge to the test solution; contact is made through an asbestos fiber. A silver-silver chloride reference electrode contains a silver wire, connected to a silver ring coated with silver chloride, immersed in a potassium chloride solution which has been saturated with silver chloride. A porous ceramic disc is used for electrical contact with the test solution. Both types of reference electrodes have a band-covered fill hole to replace the internal electrolyte which slowly leaks out the tip of the electrode as it is being used.

Limitations of pH Meter Measurements

Several limitations to the measurement of pH are known to exist. These are summarized below.

1. *The alkaline error.* An ordinary glass electrode becomes somewhat sensitive to alkali metals at pH values greater than 10. There is a tendency for pH measurements above 10 to be low, sometimes being as much as 1.0 pH unit low. Specially constructed electrodes designed for high pH measurements are available which greatly reduce this error.

2. *The acid error.* At pH values near 0, readings tend to be somewhat high.

3. *Unbuffered neutral solutions.* Equilibrium between the electrode surface and the solution is obtained very slowly in poorly buffered solutions near a pH of 7. A few minutes of constant stirring should be allowed before taking the reading.

4. *Dehydration.* When the membrane of the glass electrode becomes dehydrated, hydrogen ions cannot quickly establish equilibrium, and readings become erratic. Glass electrodes should be allowed to soak in distilled water when not in use.

5. *Variation in junction potential.* In the equation relating the potential of the glass electrode to the pH, the constant K was said to contain an asymmetry potential. Calibration of the pH meter is designed to eliminate any error introduced by the asymmetry potential. Also included in K were reference electrode potentials and junction potentials between the reference electrodes and their respective solutions for both the internal silver-silver chloride electrode in the glass electrode and the SCE or silver-silver chloride reference electrode. Junction potentials result from the unequal diffusion of ions on each side of a liquid boundary. The assumption is made that K will have the same value with the test solution as with the buffer. This is not strictly true because the junction potential varies slightly with the composition of the solution being measured. This uncertainty amounts to about $\pm$ 1 mv or ±0.02 pH unit.

6. *Temperature variations.* At 25°C one pH unit corresponds to 0.059 volts. The pH meter is designed to register a change of 1 pH unit for each potential change of 0.059 volts. At temperatures higher than 25°C, a larger potential difference corresponds to a change of 1 pH unit. At temperatures below 25°C, a smaller difference is observed. A temperature control compensator is provided to eliminate this error.

7. *Error in pH of the buffer solution.* The pH meter is calibrated with a buffer solution. Any errors in preparation or changes in composition during storage will be reflected as errors in pH measurements. Errors also can be minimized by using a buffer to calibrate the pH meter at a pH value fairly close to that of the test solution.

Questions

1. Define the following terms and give an example of each.
   (a) reducing agent   (b) half reaction
   (c) oxidation-reduction couple (d) standard electrode potential
   (e) indicator electrode   (f) reference electrode
   (g) Nernst equation   (h) potentiometric titration

2. Distinguish between a galvanic cell and an electrolytic cell.

3. Write a shorthand cell representation for a cell consisting of a silver electrode in a $0.015\,M$ silver nitrate solution connected by a potassium nitrate salt bridge to a $0.25\,M$ nickel chloride solution containing a nickel electrode.

4. What factors must be considered to determine the proper indicator for an oxidation-reduction titration?

5. What is the function of a salt bridge?

6. If the saturated calomel electrode were adopted as the standard reference electrode and assigned a potential of 0.00 volts, what would be the standard potentials of the following couples?

$$2H^+ + 2e \rightleftharpoons H_2$$

$$Ag^+ + e \rightleftharpoons Ag$$

$$Cr_2O_7^= + 14H^+ + 6e \rightleftharpoons 2Cr^{+++}$$

7. List the following substances in order of decreasing strengths as reducing agents: Fe^{++}, Sn^{++}, Hg_2^{++}, Ag.

8. Explain why a pH meter must be standardized prior to use.

9. Why does an ordinary glass electrode give erroneous results at a high pH?

Problems

1. Balance the following redox equations:
   (a) $H_2SO_4 + HBr \rightarrow SO_2 + Br_2 + H_2O$
   (b) $I_2 + HNO_3 \rightarrow HIO_3 + NO + H_2O$
   (c) $PH_3 + Cu(OH)_2 \rightarrow P_4 + H_2O + Cu$
   (d) $As_2O_3 + Zn + H_2SO_4 \rightarrow AsH_3 + ZnSO_4 + H_2O$
   (e) $CuSCN + KMnO_4 + H_2SO_4$
       $\rightarrow MnSO_4 + KCN + CuSO_4 + K_2SO_4 + H_2O$
   (f) $KSCN + Al + HCl \rightarrow KCl + AlCl_3 + NH_4Cl + C + H_2S$
   (g) $C_7H_7O_3N + H_2SO_4 \rightarrow CO_2 + (NH_4)_2SO_4 + SO_2 + H_2O$
   (h) $Cr_2O_7^= + H_2S + H^+ \rightarrow Cr^{+++} + S + H_2O$
   (i) $MnO_4^- + Cl^- + H^+ \rightarrow Mn^{++} + Cl_2 + H_2O$
   (j) $IO_3^- + I^- + H^+ \rightarrow I_2 + H_2O$

2. Calculate the percent of iron in a sample weighing 0.8025 g which after reduction to Fe^{++} requires 39.25 ml of $0.1000\,N$ $KMnO_4$ for titration.

3. A chemist needs to make many oxidations in acid solution and has at his disposal potassium dichromate which sells at \$3.63 per pound and potassium permanganate which sells for \$4.35 per pound. Calculate the cost of each reagent per equivalent of oxidizing capacity.

4. Calculate the weight of potassium dichromate required to prepare 500.0 ml of 0.1000N solution.

5. Potassium permanganate solution was standardized against 0.1025 g pure sodium formate, $NaCHO_2$, in a neutral solution requiring 42.50 ml of the $KMnO_4$ according to the equation

$$3CHO_2^- + 2MnO_4^- + H_2O \rightarrow 2MnO_2 + 3CO_2 + 5OH^-$$

From this same $KMnO_4$ solution 32.30 ml was used to oxidize the divalent iron obtained from a 0.4880 g sample of ore to trivalent iron in a strong H_2SO_4 solution. Calculate the percent iron in the sample.

6. Using Table 9-2 determine if the following reactions should be spontaneous.
   (a) $Ni^{++} + 2Fe^{++} \rightarrow Ni + 2Fe^{+++}$
   (b) $2MnO_4^- + 6H^+ + 5H_3AsO_3 \rightarrow 2Mn^{++} + 5H_3AsO_4 + 3H_2O$
   (c) $Fe^{+++} + Ag + Cl^- \rightarrow Fe^{++} + AgCl(s)$
   (d) $I_2 + 2Ag \rightarrow 2I^- + 2Ag^+$
   (e) $Cu^{++} + Zn \rightarrow Cu + Zn^{++}$

7. Calculate the standard cell potential for each spontaneous reaction in problem 6.

8. Calculate the potential of the following electrodes:
   (a) platinum electrode in 0.25M Fe^{+++} and 0.10M Fe^{++}
   (b) copper electrode in 1×10^{-4}M $CuSO_4$.
   (c) platinum electrode in 0.08M MnO_4^-, 0.015M Mn^{++}, and 2.0M H^+.

9. Calculate the potential of a silver electrode immersed in a saturated solution of silver chromate. K_{sp} for $Ag_2CrO_4 = 1.1 \times 10^{-12}$.

10. Calculate the solubility product constant for silver iodide from the following data: A silver indicating electrode and a saturated calomel electrode were immersed in a saturated solution of silver iodide containing 0.1M excess iodide ion. The cell potential was 0.336 v with the SCE being more positive than the silver electrode. E(SCE) = 0.246 v.

11. Calculate the standard potential of the dichromate-chromium(III) half reaction at (a) pH = 2 and (b) pH = 5.

12. Calculate the potential of the following cells:
    (a) $Ag(s) \mid Ag^+ (0.05M) \parallel Cu^{++} (0.20M) \mid Cu(s)$
    (b) $Pt(s) \mid Fe^{+++} (0.10M), Fe^{++} (0.50M) \parallel I_3^- (I_2)0.20M, I^- 1.0M \mid Pt(s)$

13. Calculate the solubility product constant for $Mn(OH)_2$ from the following data:

$$Mn^{++} + 2e^- \rightleftharpoons Mn \qquad\qquad E^\circ = -1.18\,v$$

$$Mn(OH)_2 + 2e^- \rightleftharpoons Mn + 2OH^- \qquad E^\circ = -1.55\,v$$

14. The half cell

$$Pt, H_2(latm) \mid HX\,(0.15M),\, NaX\,(0.20M)$$

has a potential of -0.328 volts vs. SHE. Calculate the ionization constant for the weak acid HX.

15. The solubility product constant for PbS is 3×10^{-28}. Calculate $E°$ for the reaction

$$PbS + 2e^- \rightarrow Pb + S^=$$

using the data

$$Pb^{++} + 2e^- \rightarrow Pb \qquad E° = -0.126 \text{ v}$$

16. From $E°$ values calculate K_{eq} for the reaction

$$Ni^{++} + Cd \rightleftharpoons Cd^{++} + Ni$$

17. Calculate K_{eq} for the reaction

$$IO_3^- + 5I^- + 6H^+ \rightleftharpoons 3I_2 + 3H_2O$$

18. A 50.0 ml sample of $0.10M$ Fe^{++} is titrated with $0.02M$ KMnO$_4$ in a solution buffered at pH 1.00. Derive the theoretical titration curve. Calculate the potential vs. SCE at 10.0, 25.0, 40.0, 49.0, 49.9, 50.0, 50.1, and 60.0 ml of titrant.

19. Calculate the end point potential in the titration of vanadium(V) with iron(II).

$$VO_2^+ + Fe^{++} + 2H^+ \rightarrow VO^{++} + Fe^{+3} + H_2O$$

Assume a pH 1.00.

10

Precipitation Titrations

Introduction

Classical gravimetric methods have been developed for the analysis of most ions. Gravimetric methods are time-consuming and require careful handling to prevent errors. In a precipitation titration, the volume of a standard solution of a precipitating agent required to react stoichiometrically equivalent to the species being determined is measured. This is a much faster analytical method but its applications are more limited.

Requirements for Precipitation Titrations

To be suitable for a precipitation titration a chemical reaction must satisfy the three general requirements for a titration. First, the attainment of equilibrium between the precipitating agent and the species being determined must be rapid throughout the course of the titration.

This is unnecessary in a classical gravimetric procedure because an excess of precipitating agent is added and a digestion period occurs. Second, the reaction must be quantitative and should proceed by a definite stoichiometry. Again a gravimetric procedure can use an excess of precipitating agent to force the reaction to be quantitative and an impure precipitate can be purified by digestion, reprecipitation, washing, or ignition to a more pure form. The third requirement is a method must be available for locating the equivalence point of the titration.

Precipitation titrations are, therefore, somewhat limited in scope. Titrations involving silver salts are the most important and several methods for the determination of halide ions with standard silver nitrate solution are known. The precipitation reaction in each method is the same but the detection of the end point differs.

Potentiometric Titrations with Silver Nitrate

A standard solution of silver nitrate can be used for a potentiometric titration of chloride, bromide, iodide, or thiocyanate ions. A silver indicating electrode is used to follow the change in silver ion concentration (activity) during the titration.

Consider the titration of 50.0 ml of $0.100M$ NaCl with $0.100M$ $AgNO_3$. K_{sp} for AgCl = 1.8×10^{-10}. Typical calculations are shown.

At 0.0 ml titration:

$$[Cl^-] = 0.100M$$

We will define pCl as the negative logarithm of the chloride ion concentration (activity).

$$pCl = 1.0$$

$$[Ag^+] = 0$$

$$pAg = \text{indeterminate}$$

The potential of the indicating electrode E_{Ag} is also indeterminate.

At 10.0 ml titration, the chloride ion concentration is equal to the unreacted sodium chloride. The exact concentration is given by

$$[Cl^-] = \frac{40.0 \text{ ml} \times 0.100M}{60 \text{ ml}} + \frac{K_{sp(AgCl)}}{[Ag^+]}$$

The second term is negligible and can be ignored. Therefore,

$$[Cl^-] = 0.067M$$

$$pCl = 1.17$$

The silver ion concentration is controlled by the solubility product constant for silver chloride and the chloride ion concentration.

$$[Ag^+] = \frac{K_{sp}}{[Cl^-]}$$

$$= \frac{1.8 \times 10^{-10}}{0.067}$$

$$= 2.7 \times 10^{-9}$$

$$pAg = 8.57$$

The observed potential (vs. SHE) is given by

$$E = E° + \frac{0.059}{1} \log[Ag^+]$$

$$= 0.800 - (0.059)(pAg)$$

$$= 0.294 \text{ v}$$

At 50.0 ml titration, which is at the equivalence point, neither chloride ion nor silver ion is in excess.

$$[Ag^+] = [Cl^-] = \sqrt{K_{sp}}$$

$$[Cl^-] = \sqrt{1.8 \times 10^{-10}} = 1.34 \times 10^{-5}$$

$$pCl = 4.87$$

$$[Ag^+] = 1.34 \times 10^{-5}$$

$$pAg = 4.87$$

$$E_{Ag} = 0.512 \text{ v}$$

At 60.0 ml titration, the silver ion is in excess and can be calculated directly.

$$[Ag^+] = \frac{10.0\,ml \times 0.100M}{110.0\,ml}$$

$$= 9.1 \times 10^{-3}$$

$$pAg = 2.04$$

Now the chloride is calculated from the solubility product constant and the known silver ion concentration.

$$[Cl^-] = \frac{1.8 \times 10^{-10}}{9.1 \times 10^{-3}} = 2.0 \times 10^{-8}$$

$$pCl = 7.70$$

$$E_{Ag} = 0.680\,v$$

Additional titration points are given in Table 10-1. The titration curve is shown in Figure 10-1 and Figure 10-2. Figure 10-1 shows the change in pCl during the titration. Figure 10-2 shows the change in pAg and E_{Ag} during the titration with pAg values shown on the left and E_{Ag} values shown on the right.

Table 10-1

Data for Titration of 50.0 ml of 0.100M Chloride with 0.100M Silver Ion

Volume AgNO$_3$, ml	[Cl$^-$]	pCl	[Ag$^+$]	pAg	E_{Ag}, volts vs. SHE
0	1×10^{-1}	1.00	—	—	—
2	9.2×10^{-2}	1.03	2.0×10^{-9}	8.70	0.287
10	6.7×10^{-2}	1.17	2.7×10^{-9}	8.57	0.294
25	3.3×10^{-2}	1.48	5.4×10^{-9}	8.26	0.312
40	1.1×10^{-2}	1.95	1.6×10^{-8}	7.79	0.340
45	5.3×10^{-3}	2.28	3.4×10^{-8}	7.50	0.359
49	1.01×10^{-3}	3.00	1.8×10^{-7}	6.75	0.402
49.5	5×10^{-4}	3.30	3.6×10^{-7}	6.44	0.420
49.9	1×10^{-4}	4.00	1.8×10^{-6}	5.75	0.461
50.0	1.34×10^{-5}	4.87	1.34×10^{-5}	4.87	0.512
50.1	1.8×10^{-6}	5.74	1×10^{-4}	4.00	0.564
50.5	3.6×10^{-7}	6.44	5.0×10^{-4}	3.30	0.605
51	1.8×10^{-7}	6.74	9.9×10^{-4}	3.00	0.623
55	3.75×10^{-8}	7.43	4.8×10^{-3}	2.32	0.663
60	2.0×10^{-8}	7.70	9.1×10^{-3}	2.04	0.680
75	1.4×10^{-8}	7.86	1.3×10^{-2}	1.88	0.689
100	7.2×10^{-9}	8.14	2.5×10^{-2}	1.60	0.705

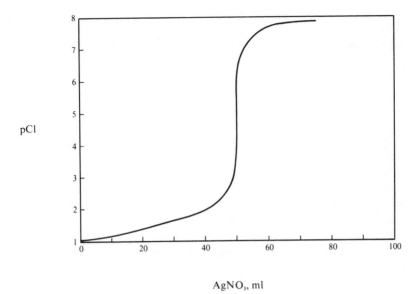

AgNO₃, ml

Figure 10-1

Titration curve for $0.100M$ NaCl with $0.100M$ AgNO₃.

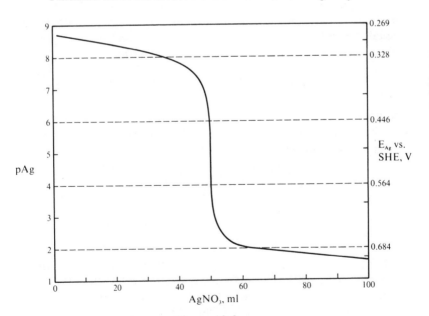

Figure 10-2

Titration curve for $0.100M$ NaCl with $0.100M$ AgNO₃.

Similar titration curves are obtained for the titration of bromide and iodide ions with silver nitrate except a larger break occurs at the equivalence points.

The Mohr Titration

The Mohr titration of chloride ion is an example of the formation of a second precipitate to indicate the completion of a precipitation titration. The second precipitate must be a different color than the first.

A small concentration of chromate ions is added as an indicator to a sample containing chloride ions. A standard solution of silver nitrate is added from a buret. The silver nitrate reacts first with the chloride ions forming the more insoluble precipitate silver chloride, which is white. After the concentration of chloride has been reduced enough, silver chromate, which is red, precipitates. The end point is taken as the first permanent darkening of the yellow chromate color of the solution. Flashes of red color will occur prior to the end point due to local excesses of titrant.

The concentration of the chromate indicator is important and should be such that the red silver chromate starts to precipitate at the equivalence point. At the equivalence point, the silver ion concentration equals the chloride ion concentration and is equal to the square root of the solubility product constant for silver chloride.

$$[Ag^+] = \sqrt{1.8 \times 10^{-10}} = 1.34 \times 10^{-5}$$

In order to have silver chromate *start* to precipitate when the chloride ion has reached this value, the chromate concentration at the equivalence point should be

$$[CrO_4^=] = \frac{K_{sp}(Ag_2CrO_4)}{[Ag^+]^2}$$

$$= \frac{1.1 \times 10^{-12}}{(1.34 \times 10^{-5})^2} = 6.1 \times 10^{-3}$$

If the chromate concentration is greater than this, the end point will come too soon. If the chromate concentration is less than this, the end point will come too late. In practice an indicator blank correction is applied because some excess silver nitrate must be added to form enough silver chromate to be seen over the white silver chloride.

The pH of the solution is an important factor in the Mohr titration. In an acid solution the following equilibria exist:

$$2CrO_4^= + 2H^+ \rightleftharpoons 2HCrO_4^- \rightleftharpoons Cr_2O_7^= + H_2O$$

Therefore, the chromate concentration is lowered in an acid solution and a larger excess of silver ion is required to cause the red precipitate to form. On the other hand, if the solution is too basic, some of the silver ion may react to form silver hydroxide:

$$Ag^+ + OH^- \rightleftharpoons AgOH_{(s)}$$

The titration should, therefore, be performed between pH 6 and pH 10.

The Mohr titration can be applied also to the titration of bromide solutions. Iodide cannot be determined in this manner because silver iodide adsorbs chromate ions preventing a distinct end point.

The Fajans Titration

The Fajans titration is an example of an adsorption indicator method for titrating chloride solutions.

Dichlorofluorescein as its sodium salt is added to a chloride solution. The solution is titrated with silver nitrate. Prior to the end point the solution contains excess chloride ions which are adsorbed as the primary adsorption layer. A cation from the solution serves as the secondary or counter adsorbed ion.

$$AgCl \cdot \cdot Cl^- \cdot \cdot \cdot \cdot M^+ \quad \text{(before equivalence point)}$$

Just past the equivalence point, silver ions are in excess and are adsorbed as the primary adsorbed layer. An anion from the solution serves as the counter adsorbed ion.

$$AgCl \cdot \cdot Ag^+ \cdot \cdot \cdot \cdot X^- \quad \text{(after equivalence point)}$$

Dichlorofluorescein is a weak organic acid, represented by HInd. When present in the titration sample the anion Ind$^-$ is preferentially adsorbed as the counter ion when there is an excess of silver ions. This gives

$$AgCl \cdot \cdot Ag^+ \cdot \cdot \cdot \cdot Ind^- \quad \text{(red)}$$

The adsorbed AgInd is red, giving a pink appearance to the precipitate. If the titration flask is swirled, the solution will appear pink.

A number of factors must be considered in selecting a precipitation titration with an adsorption indicator:

1. Since the end point occurs on the surface of the precipitate, coagulation of the precipitate should be avoided. A dilute dextrin solution acts as a protective colloid and keeps the silver chloride precipitate in a finely divided state until the end point has been reached.
2. The precipitate must strongly adsorb its own ions. This is usually the case.
3. The dye must be strongly adsorbed as a counter ion but must not be adsorbed strongly enough to displace the common ion of the precipitate. Dichlorofluorescein is used for titrations of chloride ion with silver nitrate solution. Eosin (tetrabromofluorescein) cannot be used in a chloride titration because it is too strongly adsorbed. It can be used though for the titration of bromide ion and iodide ion with silver nitrate solution.
4. Most adsorption indicators are anions of weak organic acids and must be used at a pH where sufficient anion exists for adsorption to occur. Dichlorofluorescein can be used in the pH range of 4–10. Below pH 4 insufficient anion exists for adsorption. Above pH 10 precipitation of silver hydroxide occurs. A few cationic adsorption indicators are known that can be used in strong acid solutions.
5. The indicator ion should be of opposite charge as the titrant ion. Then adsorption will not occur until excess titrant is present. Titrations with silver ion employ anionic adsorption indicators; titrations of silver ion with halide employ cationic adsorption indicators.

The Classical Volhard Titration

The Volhard method is based on the precipitation of silver thiocyanate in nitric acid solution using ferric ion to detect an excess of thiocyanate. The principle equations involved are

$$Ag^+ + SCN^- \rightleftharpoons AgSCN_{(s)}$$
$$Fe^{+++} + SCN^- \rightleftharpoons Fe(SCN)^{++}$$

The ferric-thiocyanate complex ion is deep red and appears when thiocyanate is in slight excess.

The Volhard method can be used for the direct titration of silver ion but is more often used as an indirect titration of chloride and other anions which form precipitates with silver ion. In the indirect determination, a measured excess of silver nitrate is added, and the amount of excess is determined by a back-titration with a standard solution of potassium thiocyanate. The reactions for the chloride determination are

$$Ag^+ + Cl^- \rightleftharpoons AgCl_{(s)} + Ag^+ \qquad \text{excess titration reaction}$$

$$SCN^- + Ag^+ \rightleftharpoons AgSCN_{(s)} \qquad \text{back titration reaction}$$

$$SCN^- + Fe^{+++} \rightleftharpoons Fe(SCN)^{++} \qquad \text{indicator reaction}$$

All equilibria lie far to the right. The silver chloride and silver thiocyanate precipitates are white, thus the appearance of the red complex is fairly sharp. Some practice is required for the titration because the rather high ferric concentration required imparts a slight yellow color to the solution. The solution must also be shaken vigorously near the end point because the silver thiocyanate precipitate adsorbs silver ions on its surface thus inhibiting the rate at which they combine with the thiocyanate.

In the determination of bromide and iodide, the silver halide precipitate is more insoluble than silver thiocyanate. A direct back-titration with thiocyanate can therefore be performed. However, silver chloride is *more* soluble than silver thiocyanate. Therefore, the following reaction will occur to a slight extent.

$$AgCl_{(s)} + SCN^- \rightarrow AgSCN_{(s)} + Cl^-$$

This causes low results because the volume of thiocyanate used in the back-titration will be too high. In the original Volhard chloride method, this problem was eliminated by removal of the silver chloride by filtration. The major disadvantage to filtration is that it is time-consuming. A widely used modification is that of Caldwell and Moyer which consists of coating the silver chloride precipitate with nitrobenzene prior to the back titration.* This is done by shaking the solution with a few

---

*J. R. Caldwell and H. V. Moyer, *Ind. Eng. Chem., Anal. Ed.,* **7,** 38 (1935).

milliliters of nitrobenzene thus effectively removing the precipitate from the solution without the necessity of a filtration.

The classical Volhard method can be applied to analysis of such anions as oxalate, carbonate, and arsenate, the silver salts of which are insoluble in a neutral solution but soluble in an acid solution. An excess of standard silver nitrate is added to a neutral solution of these ions. The precipitate is removed by filtration and the excess silver ion is determined in the filtrate by titration with thiocyanate after the addition of ferric ion indicator and nitric acid.

The Swift Modification of the Volhard Chloride Titration

A modification to the classical Volhard determination developed by Swift allows the direct titration of chloride ion in an acid solution.* In this method a known small amount (1.0 ml of 0.01N) of standard potassium thiocyanate is added to the sample *prior* to the titration with silver nitrate. Silver thiocyanate is more insoluble than silver chloride, so a rather strong ferric ion solution (0.2M at the end point) is added to prevent the silver thiocyanate from precipitating too soon. The reaction involved is

$$Fe^{+++} + SCN^- \rightleftharpoons Fe(SCN)^{++}$$

The equilibrium lies to the right, and with the large excess ferric ion concentration present the equilibrium concentration of thiocyanate is too small to allow the formation of silver thiocyanate until the chloride ion has been quantitatively precipitated. After the precipitation of chloride ion is complete, the silver ion reacts with the thiocyanate ion in equilibrium with the ferric-thiocyanate complex forming silver thiocyanate. When all the thiocyanate has been removed from the complex, the red color disappears. The milliequivalents of thiocyanate originally added are subtracted from the milliequivalents of silver nitrate added to give the net number of milliequivalents of chloride ion present in the sample.

The Swift modification eliminates the need for two burets but still requires two standard solutions. The volume at the end point is somewhat critical and excessive rinsing of the titration flask is to be avoided.

---

*E. H. Swift, G. M. Arcand, R. Lutwack, and D. J. Meier, *Anal. Chem.* **22**, 306 (1950).

Summary of Chloride Titrations

Of the three indicator methods for chloride (Mohr, Fajans, and Volhard), only the Volhard titration can be performed in a strong acid medium. All three methods have an inherent indicator error, so an indicator blank should be applied or the silver nitrate solution should be standardized against primary standard sodium chloride by whichever method is to be employed in the analysis of the sample.

Questions

1. What requirements limit the application of a classical gravimetric analysis to a precipitation titration procedure?

2. Explain why the Mohr titration of chloride must be performed at a pH between 6 and 10.

3. Explain why the Fajans titration of chloride must be performed at a pH between 4 and 10.

4. Why must the silver chloride be removed by filtration or coated with nitrobenzene in the classical Volhard titration of chloride?

5. Explain with the use of chemical equilibria why the Swift modification of the Volhard titration of chloride eliminates the need to effectively remove the silver chloride precipitate.

6. What is the purpose of dextrin in the Fajans titration of chloride?

7. Explain whether the failure to filter the silver chloride or to add nitrobenzene in the classical Volhard titration of chloride would cause the results to be high or low.

Problems

1. A $AgNO_3$ solution is standardized against 0.2514 g of pure NaCl by the Fajans method requiring 43.10 ml to reach the end point. Calculate the normality of the solution.

2. A 0.2885 g sample of a soluble chloride salt was analyzed by the Volhard method. To this sample was added 50.0 ml of $0.1022N$ $AgNO_3$, and the excess required 7.26 ml of KSCN. In a separate titration, 25.0 ml of the $AgNO_3$ required 31.22 ml of the KSCN. Calculate the percent chloride in the sample.

3. A silver dime containing 90.0% silver weighing 2.500 g was dissolved in nitric acid, the solution diluted to 250.0 ml and a 50.0 ml aliquot titrated

with potassium thiocyanate by the Volhard method requiring 42.30 ml to reach the end point. Calculate the normality of the KSCN solution.

4. Calculate the [Br$^-$] to [SCN$^-$] ratio in a solution saturated with silver bromide and silver thiocyanate. K_{sp}(AgBr), 5.2 $\times$ 10^{-13}; K_{sp}(AgSCN), 1.1 $\times$ 10^{-12}.

5. Calculate the bromide ion concentration in a 0.010 M solution of potassium chromate which is in equilibrium with solid AgBr and solid Ag$_2$CrO$_4$. K_{sp}(AgBr), 5.2 $\times$ 10^{-13}; K_{sp}(Ag$_2$CrO$_4$), 2.4 $\times$ 10^{-12}.

6. A 0.10N solution of sodium iodate is titrated with 0.10N silver nitrate using chromate ion as an indicator. The chromate ion concentration at the end point is 2.0 $\times$ 10^{-3}M. K_{sp}(AgI), 8.3 $\times$ 10^{-17}; K_{sp}(Ag$_2$CrO$_4$), 2.4 $\times$ 10^{-12}. Calculate the iodate concentration at the end point.

7. Calculate the pCl and pAg of a solution prepared by mixing 40.0 ml of 0.10N NaCl and 50.0 ml of 0.125N AgNO$_3$.

8. A 50.0 ml sample of 0.10N NaI is titrated with 0.10N silver nitrate. Calculate the pI after the addition of 49.9 and 50.1 ml of titrant. Compare the change in pI to the change in pCl (Table 10-1) for the same volumes. Interpret this difference in terms of the solubility of the two silver salts. K_{sp}(AgI), 8.3 $\times$ 10^{-17}; K_{sp}(AgCl), 1.8 $\times$ 10^{-10}.

9. A 50.0 ml sample of a solution which is 0.05N in NaCl and 0.05N in KI is titrated with 0.10N AgNO$_3$. Calculate and plot the pAg after the addition of 10.0, 20.0, 24.0, 24.9, 25.0, 25.1, 26.0, 35.0, 45.0, 49,0, 49.9, 50.0, 50.1, 51.0 and 60.0 ml of titrant.

10. Explain why the change in pAg obtained in problem 8 is symmetrical around the equivalence point, whereas the change in problem 9 about the first equivalence point is not symmetrical.

11

Complexometric Titrations

Introduction

Metal ions can act as electron-pair acceptors, reacting with electron donors, or ligands, to form complex ions or coordination compounds. The ligand must have at least one pair of unshared electrons to form the coordination bond. Simple ligands that form complexes with many ions are ammonia, water, chloride ion, and cyanide ion.

A given metal ion normally forms a maximum of two, four, or six coordination bonds. The number of such bonds formed is known as the coordination number of the metal. The complex species formed may be a cation, an anion, or a neutral molecule.

A ligand which has but a single pair of unshared electrons available for coordination is known as a monodentate or an unidentate. Many complexes are formed, however, where the ligand has two or more donor groups. This type of complex is known as a chelate (after the Greek word *chele* meaning claw) and the ligand is known as a chelating agent. A chelating agent containing two groups which form coordination bonds is a bidentate; if it forms three bonds it is a terdentate.

257

Other chelating agents are known as quadridentate, quinquedentate, and sexidentate. The word dentate is derived from the Latin word for tooth.

Choice of Titrant

For a complexometric titration to be suitable, it must fulfill the same requirements as all other volumetric methods. It must be rapid, proceed by a known stoichiometry, and the end point must be detectable. Many metal complexes which are very stable form too slowly for a direct titration to be suitable. In these cases, it may be possible to add an excess of the titrant, and after equilibrium has been established, the excess titrant can be determined by a back-titration.

Unidentates react to form complexes in a stepwise manner. For example the reaction of copper(II) ion with ammonia can be written

$$Cu^{++} + 4NH_3 \rightleftharpoons Cu(NH_3)_4^{++}$$

but it actually proceeds one step at a time.

$$Cu^{++} + NH_3 \rightleftharpoons Cu(NH_3)^{++}$$

$$K_1 = \frac{[Cu(NH_3)^{++}]}{[Cu^{++}][NH_3]} = 10^{4.1}$$

$$Cu(NH_3)^{++} + NH_3 \rightleftharpoons Cu(NH_3)_2^{++}$$

$$K_2 = \frac{[Cu(NH_3)_2^{++}]}{[Cu(NH_3)^{++}][NH_3]} = 10^{3.5}$$

$$Cu(NH_3)_2^{++} + NH_3 \rightleftharpoons Cu(NH_3)_3^{++}$$

$$K_3 = \frac{[Cu(NH_3)_3^{++}]}{[Cu(NH_3)_2^{++}][NH_3]} = 10^{2.9}$$

$$Cu(NH_3)_3^{++} + NH_3 \rightleftharpoons Cu(NH_3)_4^{++}$$

$$K_4 = \frac{[Cu(NH_3)_4^{++}]}{[Cu(NH_3)_3^{++}][NH_3]} = 10^{2.1}$$

The overall equilibrium constant for the complex $Cu(NH_3)_4^{++}$ is

$$K = \frac{[Cu(NH_3)_4^{++}]}{[Cu^{++}][NH_3]^4} = 10^{12.6}$$

which would indicate a stable complex ion. An end point cannot be detected in the titration of copper(II) with ammonia because the complex forms in a stepwise manner. Rarely are reactions involving unidentates suitable for complexometric titrations.

A chelating agent can occupy all the coordination sites of a particular metal ion thereby eliminating the problem of stepwise complex formation. The stability of many of these 1:1 metal-ligand complexes is high which enables a precise location of the end point.

EDTA as a Titrant

The most important titrant for complexometric titrations is EDTA (ethylenediaminetetraacetic acid).

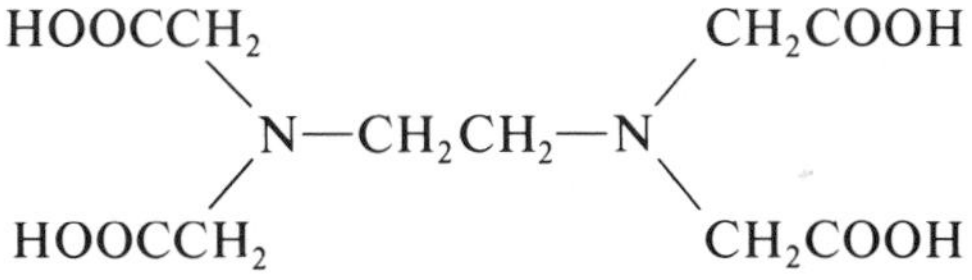

$$HOOCCH_2 \quad\quad CH_2COOH$$
$$N-CH_2CH_2-N$$
$$HOOCCH_2 \quad\quad CH_2COOH$$

EDTA

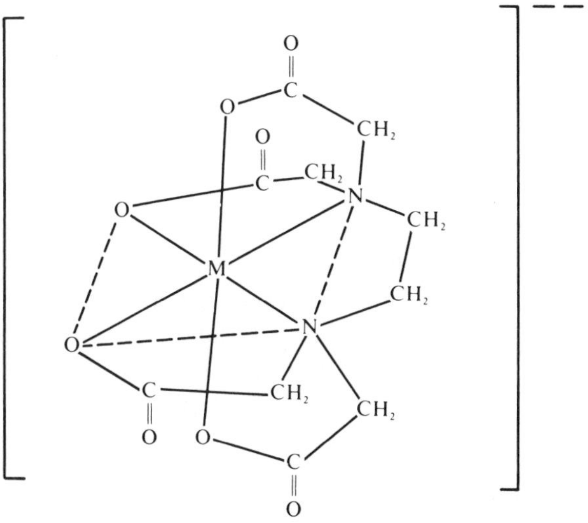

Figure 11-1

A typical metal-EDTA chelate.

EDTA forms stable complexes with many metal ions. It is a sexidentate, forming coordination bonds through its four carboxylate
groups and two nitrogen atoms. This is shown in Figure 11-1. In some
complexes EDTA acts as a quinquedentate or quadridentate having
one or two of its carboxylate groups free of strong interaction with
the metal ion. Regardless of the number of coordination groups used,
EDTA always reacts with metal ions in a 1:1 molar ratio. All metal-
EDTA complexes are water soluble and most are colorless or only
slightly colored.

EDTA is often written as H_4Y in order to show the tetraprotic behavior of the free acid. The four stepwise dissociations and their constants are

$$H_4Y \rightleftharpoons H^+ + H_3Y^-$$

$$K_1 = \frac{[H^+][H_3Y^-]}{[H_4Y]} = 8.5 \times 10^{-3} \qquad (11\text{-}1)$$

$$H_3Y^- \rightleftharpoons H^+ + H_2Y^=$$

$$K_2 = \frac{[H^+][H_2Y^=]}{[H_3Y^-]} = 1.8 \times 10^{-3} \qquad (11\text{-}2)$$

$$H_2Y^= \rightleftharpoons H^+ + HY^{-3}$$

$$K_3 = \frac{[H^+][HY^{-3}]}{[H_2Y^=]} = 5.8 \times 10^{-7} \qquad (11\text{-}3)$$

$$HY^{-3} \rightleftharpoons H^+ + Y^{-4}$$

$$K_4 = \frac{[H^+][Y^{-4}]}{[HY^{-3}]} = 4.6 \times 10^{-11} \qquad (11\text{-}4)$$

The Y^{-4} ion represents the ethylenediaminetetraacetate ion.

The free acid (H_4Y) and the monosodium salt (NaH_3Y) forms of
EDTA are only sparingly soluble in water. The disodium salt, com-
mercially available as $Na_2H_2Y \cdot 2H_2O$, is normally used to prepare
standard solutions; the salt either being weighed as a primary standard
or the solution standardized by titration of a standard solution of a
metal ion.

The distribution of EDTA among its different forms varies considerably with pH. The reactive EDTA species is Y^{-4} rather than
$H_2Y^=$, and the general reaction for the complex formation may be
written as

$$M^{+n} + Y^{-4} \rightleftharpoons MY^{n-4}$$

An increase in the acidity of the solution weakens the complex by protonating the Y^{-4} and reducing its availability for complexation as shown in the equation

$$MY^{n-4} + 2H^+ \rightleftharpoons M^{+n} + H_2Y^=$$

Consequently, most EDTA titrations are performed in a buffered neutral or slightly basic solution.

The quantitative effect of hydrogen ions on the equilibrium can be calculated with the use of β_{Y-4} which is defined as the fraction of all forms of uncomplexed EDTA present as Y^{-4}.

$$\beta_{Y-4} = \frac{[Y^{-4}]}{[H_4Y] + [H_3Y^-] + [H_2Y^=] + [HY^{-3}] + [Y^{-4}]} \qquad \textbf{(11-5)}$$

To evaluate β_{Y-4} as a function of pH requires the inverse of Equation (11-5).

$$\frac{1}{\beta_{Y-4}} = \frac{[H_4Y]}{[Y^{-4}]} + \frac{[H_3Y^-]}{[Y^{-4}]} + \frac{[H_2Y^=]}{[Y^{-4}]} + \frac{[HY^{-3}]}{[Y^{-4}]} + \frac{[Y^{-4}]}{[Y^{-4}]} \qquad \textbf{(11-6)}$$

The last term of Equation (11-6) is unity and the other terms may be evaluated from the ionization constants for EDTA. The next to last term in Equation (11-6) can be obtained from Equation (11-4).

$$\frac{[HY^{-3}]}{[Y^{-4}]} = \frac{[H^+]}{[K_4]} \qquad \textbf{(11-7)}$$

Rearrangement of Equation (11-3) gives

$$[H_2Y^=] = \frac{[H^+][HY^{-3}]}{K_3} \qquad \textbf{(11-8)}$$

Solving Equation (11-7) for $[HY^{-3}]$ and substitution of it into Equation (11-8) gives

$$[H_2Y^=] = \frac{[H^+][H^+][Y^{-4}]}{K_3K_4}$$

or

$$\frac{[H_2Y^=]}{[Y^{-4}]} = \frac{[H^+]^2}{K_3 K_4} \qquad (11\text{-}9)$$

Equation (11-9) is the third term of Equation (11-6).
  Rearrangement of Equation (11-2) gives

$$[H_3Y^-] = \frac{[H^+][H_2Y^=]}{K_2} \qquad (11\text{-}10)$$

Solving Equation (11-9) for $[H_2Y^=]$ and substitution of it into Equation (11-10) gives

$$[H_3Y^-] = \frac{[H^+][H^+]^2[Y^{-4}]}{K_2 K_3 K_4}$$

or

$$\frac{[H_3Y^-]}{[Y^{-4}]} = \frac{[H^+]^3}{K_2 K_3 K_4} \qquad (11\text{-}11)$$

Equation (11-11) is the second term of Equation (11-6).
  Rearrangement of Equation (11-1) gives

$$[H_4Y] = \frac{[H^+][H_3Y^-]}{K_1} \qquad (11\text{-}12)$$

Solving Equation (11-11) for $[H_3Y^-]$ and substitution of it into Equation (11-12) gives

$$[H_4Y] = \frac{[H^+][H^+]^3[Y^{-4}]}{K_1 K_2 K_3 K_4}$$

or

$$\frac{[H_4Y]}{[Y^{-4}]} = \frac{[H^+]^4}{K_1 K_2 K_3 K_4} \qquad (11\text{-}13)$$

Equation (11-13) is the first term of Equation (11-6).

Therefore,

$$\frac{1}{\beta_{Y-4}} = \frac{[H^+]^4}{K_1 K_2 K_3 K_4} + \frac{[H^+]^3}{K_2 K_3 K_4} + \frac{[H^+]^2}{K_3 K_4} + \frac{[H^+]}{K_4} + 1 \qquad \textbf{(11-14)}$$

Depending on the pH usually only one to three terms in Equation (11-14) will be significant.

EXAMPLE 11-1: Calculate β_{Y-4} for EDTA at pH 10.0.

$$\frac{1}{\beta_{Y-4}} = \frac{10^{-40}}{K_1 K_2 K_3 K_4} + \frac{10^{-30}}{K_2 K_3 K_4} + \frac{10^{-20}}{K_3 K_4} + \frac{10^{-10}}{K_4} + 1$$

$$= 2.4 \times 10^{-19} + 2.1 \times 10^{-11} + 3.7 \times 10^{-4} + 2.17 + 1$$

$$= 3.17$$

$$\beta_{Y-4} = 0.315$$

In a manner similar to that above, expressions for β_{HY-3}, $\beta_{H_2Y^-}$, β_{H_3Y-}, and β_{H_4Y} can be derived. These expressions are

$$\frac{1}{\beta_{HY-3}} = \frac{[H^+]^3}{K_1 K_2 K_3} + \frac{[H^+]^2}{K_2 K_3} + \frac{[H^+]}{K_3} + 1 + \frac{K_4}{[H^+]}$$

$$\frac{1}{\beta_{H_2Y^-}} = \frac{[H^+]^2}{K_1 K_2} + \frac{[H^+]}{K_2} + 1 + \frac{K_3}{[H^+]} + \frac{K_3 K_4}{[H^+]^2}$$

$$\frac{1}{\beta_{H_3Y-}} = \frac{[H^+]}{K_1} + 1 + \frac{K_2}{[H^+]} + \frac{K_2 K_3}{[H^+]^2} + \frac{K_2 K_3 K_4}{[H^+]^3}$$

$$\frac{1}{\beta_{H_4Y}} = 1 + \frac{K_1}{[H^+]} + \frac{K_1 K_2}{[H^+]^2} + \frac{K_1 K_2 K_3}{[H^+]^3} + \frac{K_1 K_2 K_3 K_4}{[H^+]^4}$$

Table 11-1 consists of the calculated values for the fraction of each EDTA species at pH values from 0 to 14. Figure 11-2 shows a plot of the same data. Figure 11-3 also shows a plot of the same data except in a logarithmic form. Reference to these graphs indicates what species are present in a significant concentration at a given pH.

It is obvious from the data presented in Table 11-1 and plotted in Figures 11-2 and 11-3 that a competition occurs between the metal ion and the hydrogen ion for the complexing agent. At pH values 10 and above, a significant fraction of the EDTA exists as Y^{-4}. At lower pH values protonated species predominate.

Table 11-1

Fraction of EDTA Species Present at Different pH Values

pH	β_{H_4Y}	$\beta_{H_3Y^-}$	$\beta_{H_2Y^=}$	$\beta_{HY^{-3}}$	$\beta_{Y^{-4}}$
0	9.92×10^{-1}	8.4×10^{-3}	1.5×10^{-5}	8.9×10^{-12}	4.0×10^{-22}
1	9.20×10^{-1}	7.8×10^{-2}	1.4×10^{-3}	8.2×10^{-9}	3.8×10^{-17}
2	4.99×10^{-1}	4.24×10^{-1}	7.6×10^{-2}	4.4×10^{-6}	2.1×10^{-14}
3	4.0×10^{-2}	3.42×10^{-1}	6.17×10^{-1}	3.6×10^{-4}	1.6×10^{-11}
4	6.2×10^{-4}	5.2×10^{-2}	9.41×10^{-1}	5.5×10^{-3}	2.5×10^{-9}
5	6.1×10^{-6}	5.2×10^{-3}	9.40×10^{-1}	5.5×10^{-2}	2.4×10^{-7}
6	4.3×10^{-8}	3.5×10^{-4}	6.32×10^{-1}	3.66×10^{-1}	1.7×10^{-5}
7	9.6×10^{-11}	8.2×10^{-6}	1.47×10^{-1}	8.52×10^{-1}	3.9×10^{-4}
8	1.1×10^{-13}	9.4×10^{-8}	1.7×10^{-2}	9.78×10^{-1}	4.6×10^{-3}
9	1.1×10^{-16}	9.1×10^{-10}	1.6×10^{-3}	9.54×10^{-1}	4.4×10^{-2}
10	7.7×10^{-20}	6.6×10^{-12}	1.2×10^{-4}	6.85×10^{-1}	3.15×10^{-1}
11	2.0×10^{-23}	1.7×10^{-14}	3.1×10^{-6}	1.78×10^{-1}	8.2×10^{-1}
12	2.4×10^{-27}	2.0×10^{-17}	3.7×10^{-8}	2.1×10^{-2}	9.78×10^{-1}
13	2.4×10^{-31}	2.1×10^{-20}	3.7×10^{-10}	2.2×10^{-3}	9.98×10^{-1}
14	2.4×10^{-35}	2.1×10^{-23}	3.7×10^{-12}	2.2×10^{-4}	1.000

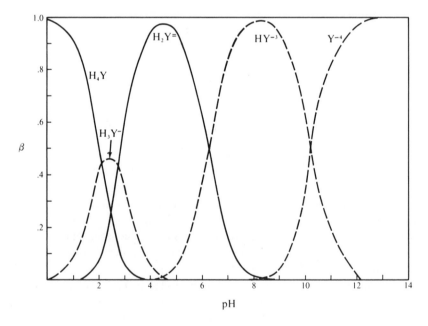

Figure 11-2

Distribution of EDTA species as a function of pH.

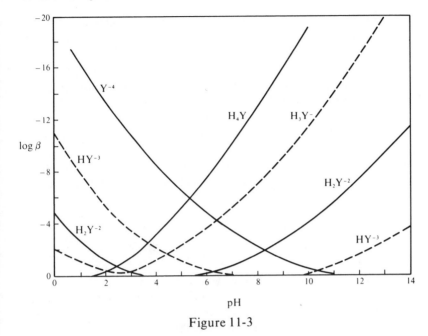

Figure 11-3

Fraction of EDTA species present as a function of pH.

Conditional Equilibrium Constants

Effect of pH

The equilibrium constant for the formation of a complex (called formation constant or stability constant) is a measure of the strength of a complex. The general reaction of a metal ion with EDTA was given previously as

$$M^{+n} + Y^{-4} \rightleftharpoons MY^{n-4}$$

Simplified, by ignoring charges, the reaction is

$$M + Y \rightleftharpoons MY$$

The formation constant expression is

$$K = \frac{[MY]}{[M][Y]} \tag{11-15}$$

Equation (11-15) is known as the absolute formation constant for the complex. Table 11-2 lists some absolute formation constants for some metal-EDTA complexes. By use of β_{Y-4}, which we will now designate as β_Y, a conditional constant can be calculated for a metal-EDTA complex at any pH. If we now define $[Y']$ as the total concentration of uncomplexed EDTA in all forms we can write

$$[Y] = \beta_Y [Y'] \qquad \textbf{(11-16)}$$

Equation (11-16) gives the concentration of EDTA available for complexation (Y^{-4}). Substituting Equation (11-16) into Equation (11-15) yields

$$K = \frac{[MY]}{[M][Y']\beta_Y} \quad \text{or} \quad K\beta_Y = \frac{[MY]}{[M][Y']} \qquad \textbf{(11-17)}$$

The term $K\beta_Y$ sometimes written as K' or K_{eff} is known as the *conditional formation constant,* so called because its value is conditional on the pH value of the solution. The larger the conditional formation constant, the stronger the complex and the sharper the inflection at the equivalence point. To perform an accurate titration with EDTA the minimum conditional formation constant should be approximately 10^8.

Table 11-2

Formation Constants of Metal-EDTA Complexes

Metal ion	log K	Metal ion	log K
Ag^+	7.3	La^{+++}	15.4
Al^{+++}	16.1	Mg^{++}	8.7
Ba^{++}	7.8	Mn^{++}	14.0
Bi^{+++}	22.8	Ni^{++}	18.6
Ca^{++}	10.7	Pb^{++}	18.0
Cd^{++}	16.5	Sr^{++}	8.6
Ce^{+++}	16.0	Th^{+4}	23.2
Co^{++}	16.3	TiO^{++}	17.3
Cr^{+++}	23.0	VO^{++}	18.8
Cu^{++}	18.8	Y^{+++}	18.1
Fe^{++}	14.3	Zn^{++}	16.5
Fe^{+++}	25.1		
Hg^{++}	21.8		

Strong metal-EDTA complexes can be formed in slightly acidic solutions, whereas weaker complexes must be formed in a basic solution to be acceptable for a complexometric titration.

EXAMPLE 11-2: Calculate the conditional formation constants for Zn-EDTA and Ca-EDTA at pH 6.0.
From Table 11-1 we see that $\beta_{Y^{-4}}$ is 1.7×10^{-5} or $10^{-4.8}$ at pH 6.0.

$$K'_{\text{Zn-EDTA}} = 10^{16.5} \times 10^{-4.8}$$

$$= 10^{11.7}$$

$$K'_{\text{Ca-EDTA}} = 10^{10.7} \times 10^{-4.8}$$

$$= 10^{5.9}$$

The titration of zinc with EDTA is feasible at pH 6.0, whereas calcium cannot be titrated at pH 6.0.

Effect of a Complexing Buffer

Just as the hydrogen ion concentration affects the availability of Y^{-4} for complexation, the presence of a buffer which forms a weak complex with the metal ion affects the availability of the metal ion for complexation. This also affects the formation constant and can also be handled in the conditional formation constant. Many EDTA titrations are performed in an ammonia-ammonium ion buffer solution. Ammonia forms a complex with many metal ions which are complexed by the EDTA. The fraction of metal ion available for complexation by the EDTA is calculated as follows:

$$\beta_M = \frac{[M_u]}{[M_t]}$$

where M_u represents the metal ion concentration uncomplexed by either ammonia or EDTA and M_t represents the total metal ion concentration as free ions and metal-ammonia complex ions but does not include the metal ions complexed by the EDTA.

If ammonia forms a series of four stepwise complexes with the metal ion, the value of β_M is calculated from Equation (11-18).

$$\frac{1}{\beta_M} = 1 + K_1[NH_3] + K_1K_2[NH_3]^2$$

$$(11\text{-}18)$$

$$+ K_1K_2K_3[NH_3]^3 + K_1K_2K_3K_4[NH_3]^4$$

where K_1, K_2, K_3, K_4 are the stepwise formation constants for the metal-ammonia (called metal-ammine) complexes and $[NH_3]$ is the concentration of uncomplexed ammonia.

Combining both β_M and β_{Y-4} with the formation constant gives a conditional formation constant valid for a given set of conditions. Again this value should be 10^8 or greater for a satisfactory titration. This is illustrated in Example 11-3.

EXAMPLE 11-3: Calculate the conditional formation constant for Zn-EDTA in a solution containing $0.1M$ NH_4^+ and $0.55M$ uncomplexed NH_3. Logarithms of the zinc-ammine complexes are K_1, 2.37; K_2, 2.44; K_3, 2.50; K_4, 2.15.

$$\frac{1}{\beta_M} = 1 + (10^{2.37})(0.55) + (10^{5.81})(0.55)^2$$

$$+ (10^{8.31})(0.55)^3 + (10^{10.46})(0.55)^4$$

$$= 1 + 129 + 1.95 \times 10^5 + 3.40 \times 10^7$$

$$+ 2.64 \times 10^9 = 2.67 \times 10^9$$

$$\beta_M = \frac{1}{2.67 \times 10^9} = 3.7 \times 10^{-10} = 10^{-9.4}$$

The pH of the buffer solution is 10.0 and β_{Y-4} (from Table 11-1) is $10^{-0.50}$.

The conditional formation constant is

$$K' = K_{Zn\text{-}EDTA}\beta_M\beta_{Y-4}$$

$$= 10^{16.5} \times 10^{-9.4} \times 10^{-.5}$$

$$= 10^{6.6}$$

The effect of a second complexing agent is often put to direct use to overcome interferences. For example, nickel forms a very stable complex ion with cyanide, $Ni(CN)_4^=$, whereas lead does not. Therefore, if cyanide is added to the solution, lead in the presence of nickel can be titrated with EDTA without any interference from the nickel; even though the formation constants for the two metal-EDTA complexes are almost identical. The effect of cyanide in this example is called *masking.*

The ammonia buffer used in many EDTA titrations, in addition to affecting β_{Y-4}, keeps the metal ion from forming a precipitate of metal hydroxide by complexing the metal ion forming an ammine. It must be remembered that the presence of ammonia affects the conditional formation constant in the titration of such metal ions with EDTA in two ways. First, by affecting the pH, it affects the availability of the complexing agent, β_{Y-4}; and second, it affects the availability of the metal ion for complexing, β_M. Both effects must be considered in calculating the conditional formation constant which must remain above about 10^8 to allow a satisfactory titration.

Calculation of a Theoretical Titration Curve

Calculation of the metal ion concentration at various stages in a titration can utilize all the principles developed in this chapter. We will consider the titration of 50.0 ml of $0.01M$ Cd^{++} containing $0.10M$ NH_4^+ and $0.10M$ NH_3 with $0.01M$ EDTA. The formation constant for Cd-EDTA is $10^{16.5}$ and the formation constants for cadmium-ammine complexes are K_1, $10^{2.60}$; K_2, $10^{2.05}$; K_3, $10^{1.39}$; K_4, $10^{0.88}$.

We will assume that at all stages in the titration prior to the end point sufficient excess ammonia is present and the cadmium exists primarily as $Cd(NH_3)_4^{++}$. As the titration proceeds, the concentration of uncomplexed ammonia changes, therefore, β_M changes.

At 0.00 ml titration, the solution contains 0.5 mmole Cd^{++} and 5.0 mmoles NH_3. The uncomplexed ammonia concentration is 3.0 mmoles in 50ml or $0.060M$. $\beta_{Cd^{++}}$ is calculated by use of Equation (11-18).

$$\beta_{Cd^{++}} = 10^{-2.72}$$

The actual free Cd^{++} concentration $[Cd_u^{++}]$ is given by

$$\beta_{Cd^{++}} = \frac{[Cd_u^{++}]}{[Cd_t^{++}]}$$

$$[Cd_u^{++}] = [Cd_t^{++}]\beta_{Cd^{++}}$$

$$= 0.01 \times 10^{-2.72}$$
$$= 10^{-4.72}$$
$$pCd = 4.72$$

At 10.0 ml titration, the total cadmium concentration uncomplexed by EDTA is $6.67 \times 10^{-3} M$. Therefore, $\beta_{Cd^{++}}$ is again calculated by the use of Equation (11-18) and $[Cd_u^{++}]$ is calculated as before.

$$\beta_{Cd^{++}} = 10^{-2.65}$$
$$pCd = 4.83$$

At 20.0 ml titration:

$$\beta_{Cd^{++}} = 10^{-2.60}$$
$$pCd = 4.97$$

At 40.0 ml titration:

$$\beta_{Cd^{++}} = 10^{-2.53}$$
$$pCd = 5.48$$

At 49.0 ml titration:

$$\beta_{Cd^{++}} = 10^{-2.51}$$
$$pCd = 6.51$$

At 49.9 ml titration:

$$\beta_{Cd^{++}} = 10^{-2.51}$$
$$pCd = 7.51$$

At the equivalence point:

$$\beta_{Cd^{++}} = 10^{-2.51}$$

The pH at the equivalence point is 9.25, therefore,

$$\beta_{Y-4} = 10^{-1.12}$$

$$K'_{Cd\,EDTA} = K_{CdEDTA}\beta_{Cd++}\beta_{Y-4}$$

$$= (10^{16.5})(10^{-2.51})(10^{-1.12})$$

$$= 10^{12.9}$$

$$K' = \frac{[CdY^=]}{[Cd_t^{++}][Y_t^{-4}]}$$

where $[Cd_t^{++}]$ is the total $[Cd^{++}]$ not complexed by EDTA and $[Y_t^{-4}]$ is the total excess uncomplexed EDTA in all species.

At the equivalence point

$$[Cd_t^{++}] = [Y_t^{-4}]$$

$$[CdY^{-2}] = 0.005\,M$$

therefore,

$$10^{12.9} = \frac{10^{-2.30}}{[Cd_t^{++}]^2}$$

$$[Cd_t^{++}] = 10^{-7.60}$$

$$[Cd_u^{++}] = [Cd_t^{++}]\beta_{Cd++}$$

$$= (10^{-7.60})(10^{-2.51})$$

$$= 10^{-10.11}$$

$$pCd = 10.11$$

At 50.1 ml titration an excess of EDTA is present.

$$[Cd_t^{++}] = \frac{[CdY^=]}{K'[Y_t^{-4}]}$$

$$[Y_t^{-4}] = 1 \times 10^{-5}\,M$$

$$[CdY^=] = 0.005\,M$$

$$K' = 10^{12.9}$$

$$[Cd_i^{++}] = \frac{0.005}{(10^{12.9})(1 \times 10^{-5})}$$

$$= 10^{-10.2}$$

$$\beta_{Cd^{++}} = 10^{-2.51}$$

$$pCd = 12.7$$

At 51.0 ml:

$$\beta_{Cd^{++}} = 10^{-2.50}$$

$$pCd = 13.7$$

At 60.0 ml:

$$\beta_{Cd^{++}} = 10^{-2.40}$$

$$pCd = 14.7$$

At 75.0 ml:

$$\beta_{Cd^{++}} = 10^{-2.26}$$

$$pCd = 15.0$$

These data are plotted in Figure 11-4.

The quantitative effect of pH and a complexing buffer can be shown by comparing the complexometric titration curve of a metal ion with a ligand, where the pH and the buffer have no effect on the availability of the two reacting species, with the curve obtained in the presence of a complexing buffer and with the curve obtained where pH has an effect. In each case we will assume a formation constant of $10^{-16.5}$ which is identical to that for Cd-EDTA. Data for these curves are given in Table 11-3 and the effects are shown in Figure 11-5.

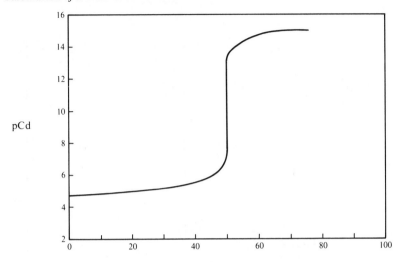

Volume EDTA, ml

Figure 11-4

Theoretical titration curve for 50.0 ml of 0.01 M
Cd^{++} with 0.01 M EDTA. The original solution
contains 0.10M NH$_4^+$ and 0.10M NH$_3$.

Table 11-3

Data for Titration of 50.0 ml of 0.01 M Metal (M) Solution
with 0.01 M Ligand (L); Formation Constant, $10^{16.50}$.

(a) No effect of pH or complexing buffer. (b) Effect of pH with
β_L, $10^{-1.1}$. (c) Effect of complexing buffer with β_M, $10^{-2.5}$.

| | pM | | |
Volume, ml	(a)	(b)	(c)
0.0	2.00	2.00	4.50
10.0	2.18	2.18	4.68
20.0	2.37	2.37	4.87
40.0	2.95	2.95	5.45
49.0	4.00	4.00	6.50
49.9	5.00	5.00	7.50
50.0	9.90	8.85	10.65
50.1	13.80	12.70	13.80
51.0	14.80	13.70	14.80
60.0	15.80	14.70	15.80
75.0	16.20	15.10	16.20

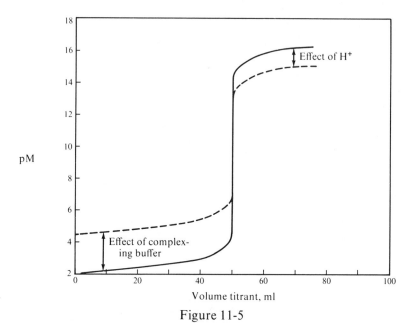

pM

Volume titrant, ml

Figure 11-5

Effect of pH and complexing buffer on a complexometric titration. The solid line represents the curve
in the absence of the complexing of the metal ion by
buffer or the complexing of the ligand by hydrogen
ion.

Use of EDTA for Determination of Water Hardness

The determination of total water hardness is a widely used application
of a complexometric titration. Total hardness is due to calcium and
magnesium ions in the water. The sum of the two ions is determined
and usually reported as calcium carbonate in either parts per million or
grains per gallon.

   The indicator used in the titration is usually Eriochrome Black T or
Calmagite both of which form complexes with both calcium and magnesium. Both indicators are tribasic acids having a sulfonic acid group
and two acidic phenolic groups in their structure. Either molecule can
be abbreviated as H_3Ind. The sulfonic acid group is a strong acid so in
aqueous solution both indicators exist as H_2Ind^- and are red in color.
The color changes which these indicators undergo with change in pH
are summarized by the following equations:

$$H_2 Ind^- \xrightleftharpoons{\text{pH = 6.3}} HInd^= \xrightleftharpoons{\text{pH = 11.6}} Ind^{-3} \qquad \textbf{(11-19)}$$
$$\quad\; \text{red} \qquad\qquad\quad \text{blue} \qquad\qquad\quad\; \text{orange}$$

$$H_2Ind^- \xrightleftharpoons{\text{pH = 8.1}} HInd^= \xrightleftharpoons{\text{pH = 12.4}} Ind^{-3} \qquad \textbf{(11-20)}$$
$$\quad\; \text{red} \qquad\qquad\quad \text{blue} \qquad\qquad\quad\; \text{orange}$$

Equation (11-19) gives the pH values at which Eriochrome Black T changes and Equation (11-20) gives the pH values at which Calmagite changes. Water hardness titrations are performed in a solution buffered at approximately pH 10 where both indicators exist as an ion which is blue.

Magnesium forms a stronger complex with the indicator than does calcium and the resulting complex is wine-red.

$$HInd^= + Mg^{++} \rightleftharpoons MgInd^- + H^+ \qquad \textbf{(11-21)}$$
$$\text{blue} \qquad\qquad\quad \text{wine red}$$

A small amount of magnesium must be present in the sample or the reaction represented in Equation (11-21) will not take place, but instead an analogous reaction with calcium will occur. Since the $CaInd^-$ complex is weaker than the $MgInd^-$ complex, it causes a premature end point and an error in the titration to occur. If no magnesium is present, a small known amount of magnesium must be added so that the correct end point will be observed. The amount of EDTA equivalent to the added magnesium must be subtracted from the total EDTA used in the titration. An alternative and preferable procedure is to add a small volume of Mg-EDTA complex before the titration. The volume of complex added does not need to be measured because a volume correction for the titrant is not necessary.

The reactions which occur in the water hardness titration using either Eriochrome Black T or Calmagite indicator in a solution buffered at pH 10 may be summarized as follows: Initially the indicator reacts with either free magnesium ions from the solution or magnesium ions from the added Mg-EDTA chelate to form a wine-red color. This reaction was represented by Equation (11-21). When EDTA is added it will react initially with the calcium ions in the solution since it forms a more stable complex with calcium than with magnesium.

$$H_2Y^= + Ca^{++} \rightleftharpoons CaY^= + 2H^+ \qquad \textbf{(11-22)}$$

When most of the calcium ions are reacted, the EDTA will begin to react with the free magnesium ions in the solution.

$$H_2Y^= + Mg^{++} \rightleftharpoons MgY^= + 2H^+ \qquad \text{(11-23)}$$

The reactions represented by Equations (11-22) and (11-23) will proceed until all the free calcium ions and magnesium ions are reacted. Then the EDTA will react with the $MgInd^-$ as shown in Equation (11-24).

$$\underset{\text{wine-red}}{H_2Y^= + MgInd^-} \rightleftharpoons \underset{\text{blue}}{MgY^= + HInd^= + H^+} \qquad \text{(11-24)}$$

When all the $MgInd^-$ has been converted into $MgY^=$ the solution changes from wine-red to blue. The end point of the titration is taken when the last trace of purple disappears at which point all the indicator is in the blue form. Obviously a large concentration of indicator is to be avoided if the end point is to be sharp.

A problem encountered with Eriochrome Black T and Calmagite is that several metal ions including iron(III), copper(II), and nickel(II) form such strong complexes with the indicator that a reaction similar to Equation (11-24) cannot occur and no end point can be observed. The presence of copper(II), nickel(II), and small amounts of iron(III) can be masked with the addition of cyanide ion.

The pH 10 ammonia-ammonium chloride buffer serves several purposes in the water hardness determination. First, at pH 10 an appreciable fraction (0.315) of the EDTA is available for complexation as Y^{-4}. Second, in order to observe an end point using Eriochrome Black T as an indicator, the pH must be between 6.3 and 11.6; using Calmagite it must be between 8.1 and 12.4. Third, hydrogen ions are produced during the reaction as seen in Equations (11-21)–(11-24). The buffer can consume the hydrogen ions as they are formed. Fourth, the use of a basic solution allows the safe use of cyanide ion for the masking of interfering ions. Many natural water supplies contain sufficient iron(III) to require the addition of cyanide ion prior to titration. The buffer must be added *prior* to the addition of cyanide ion.

The water hardness determination as described above gives the hardness due to *both* the calcium and magnesium ions. Calcium hardness only can be determined by making the solution strongly basic with sodium hydroxide which precipitates the magnesium ions as magnesium hydroxide. The calcium is then titrated with EDTA. Calcein is used as the indicator in the calcium hardness determination.

A detailed procedure for the determination of total water hardness is given in the laboratory section (Experiment 19).

Questions

1. Define the following terms: ligand, coordination number, chelate, quadridentate, conditional equilibrium constant.

2. What criteria must be met in order to have a successful complexometric titration?

3. Why is ammonia not used as a titrant for metal ions even for those with which it forms very stable complexes?

4. Draw the structure for the calcium complex with EDTA.

5. Discuss what effect pH can play on the analysis of mixtures of metal ions.

6. Explain why some magnesium must be present in a sample for a water hardness titration using Eriochrome Black T as an indicator and EDTA as a titrant.

7. In the water hardness determination using EDTA and Eriochrome Black T there are three principle complex ions. List the three complex ions in order of their increasing stability and explain how this stability can be illustrated by the results of the determination.

8. What purposes does the ammonia-ammonium chloride buffer serve in the water hardness determination with EDTA?

Problems

1. Calculate β_{Ox} for a solution of oxalic acid, H_2Ox, at pH 6.0 and pH 3.0. $K_1, 8.8 \times 10^{-2}; K_2, 5.1 \times 10^{-5}$.

2. Derive an expression for the calculation of β_{HA^-} at any pH for a solution containing H_2A, HA^-, and $A^=$.

3. A standard solution of calcium carbonate was prepared by dissolving 0.4826 g of calcium carbonate in hydrochloric acid and dilution to 100.0 ml in a volumetric flask. A 10.0 ml aliquot required 42.18 ml of EDTA for titration at pH 10. Calculate the molarity of the EDTA.

4. Assuming 10^8 to be the minimum value for a conditional formation constant for a successful titration, calculate the most acidic pH that would be acceptable for titration of magnesium with EDTA (refer to Table 11-2 and Figure 11.2).

5. Answer problem 4 for the EDTA titration of Zn^{++}. Select a suitable pH for the titration of Zn^{++} in the presence of Mg^{++}.

6. Aluminum(III) reacts too slowly with EDTA for a direct titration to be convenient. Aluminum(III) may be determined by adding an excess of standard EDTA, heating to ensure complete reaction, cooling, buffering,

and backtitrating with standard copper(II) solution. From the following
data, calculate the concentration of aluminum in a sample:

Sample volume   50.0 ml

EDTA $(0.0102M)$ 50.0 ml

Cu^{++} $(0.0202M)$ 12.1 ml

7. Construct a titration curve for the titration of 50.0 ml of $0.05M$ Ca^{++} with $0.05M$ EDTA at a pH of 10 and 8. Calculate pCa values at 0, 25, 40, 49, 49.9, 50, 50.1, 51, and 60 for each pH. Compare the feasibility of the titration at the two pH values.

8. The formation constant of $Ag(H_2NCH_2CH_2NH_2)^+$ is 5.0×10^4. Calculate the concentration of Ag^+ at equilibrium if 10 millimoles of silver nitrate is added to 25 millimoles of ethylenediamine, $H_2NCH_2CH_2NH_2$. Assume a final volume of 100 ml.

9. Using Tables 11-1 and 11-2 calculate the pM at the equivalence point for the titration of the following:
   (a) $0.01M$ Pb^{++} with $0.01M$ EDTA at pH 4.0
   (b) $0.005M$ Fe^{+++} with $0.005M$ EDTA at pH 2.0
   (c) $0.05M$ Sr^{++} with $0.05M$ EDTA at pH 8.0

10. Explain why calmagite is not a suitable indicator for the titration of Ca^{++} unless Mg^{++} is also present.

11. A solution contains calcium and magnesium. A 50.0 ml aliquot was titrated with $0.015M$ EDTA at pH 12.5 using Calcein indicator requiring 26.82 ml. A second 50.0 ml aliquot was titrated at pH 10 using Eriochrome Black T indicator requiring 30.12 ml of the EDTA solution. Calculate the parts per million of (a) calcium carbonate, (b) magnesium carbonate, and (c) total hardness *as* calcium carbonate.

12. A mixture of Zn^{++} and Mg^{++} can be analyzed by titration with EDTA with and without the addition of potassium cyanide (which complexes the Zn^{++}). A 50.0 ml aliquot of the mixture required 38.25 ml of $0.05M$ EDTA when titrated without the addition of potassium cyanide and 22.10 ml when titrated in the presence of potassium cyanide. Calculate the concentration of Zn^{++} and Mg^{++} in the sample.

12

Spectrophotometric Methods of Analysis

Introduction

Spectrophotometry involves the measurement of the absorption of radiant energy by a chemical species as a function of the wavelength of the radiation or its measurement at a given wavelength. Spectrophotometric methods are generally rapid and are adaptable to the determination of small concentrations of species. They are available for the analysis of many different types of chemical species. Instruments range from relatively cheap, manually operated instruments to sophisticated, automatic recording spectrophotometers costing thousands of dollars.

Nature of Light

Electromagnetic radiation is a form of energy which has a dual nature. Properties of light such as reflection, refraction, and diffraction are

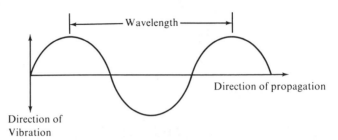

Figure 12-1

An electromagnetic wave.

best explained by the wave theory of light in which the waves travel at extreme velocities and do not require the existence of a supporting medium for propagation. Other properties such as the photoelectric effect and the absorbance of radiation by matter are best interpreted by the particle nature of light—a beam of light is a bundle of energy packets called photons.

A radiant wave vibrates perpendicular to the direction of propagation as shown in Figure 12-1.

The *wavelength* λ is the distance traveled by a complete cycle of the wave. The *frequency* ν is the number of waves passing a fixed point per unit of time. The reciprocal of the wavelength is called the *wave number* $\bar{\nu}$ and is the number of waves in a unit length. The relationship between these properties is

$$\frac{1}{\lambda} = \bar{\nu} = \frac{\nu}{c}$$

or

$$\lambda\nu = c$$

where c is the velocity of light, 2.998×10^{10} cm/sec in a vacuum and approximately the same in air.

The units commonly used for wavelength depend upon the region of the spectrum. For visible and ultraviolet radiation, the nanometer is normally used. In the infrared region the micrometer is the common unit. A micrometer, μm, is 10^{-6} m and a nanometer, nm, is 10^{-9} m. Wave numbers are often used by chemists as a frequency unit because it has convenient numerical values being related to frequency by the constant factor c, the velocity of light. Frequency is measured in cycles per second or hertz (Hz). Wave number is measured in reciprocal centimeters, cm^{-1}.

The energy of a photon is variable and depends upon the frequency or wavelength of the radiation. The relationship between the energy E of a photon and frequency ν is

$$E = h\nu$$

or

$$E = \frac{hc}{\lambda}$$

where h is *Planck's constant* with a numerical value of 6.62×10^{-27} erg sec. Therefore, as the wavelength increases the energy decreases.

The Electromagnetic Spectrum

The electromagnetic spectrum covers an immense range of wavelengths. Table 12-1 shows on a logarithmic scale the major divisions of the spectrum. The spectrum is continuous and wavelength regions are shown classified by the type of radiation. Gamma ray and X-ray regions are not considered in this text. The *ultraviolet* region extends from about 10 nm to 380 nm although the most useful region for analysis is from 200 nm to 380 nm, called the *near ultraviolet* region. Air absorbs appreciable radiation below 200 nm and this region is called the *vacuum ultraviolet* because instruments must be operated in a vacuum. The *visible* region of the spectrum is only a very small part of the electromagnetic spectrum extending from 380 nm to about 780 nm. It is this region of wavelengths which can be seen by the eye. The *infrared* region extends from 0.78 μm to 300 μm with the range 2.5 μm to 15 μm being most frequently used for analysis. Lower energy radiation (radio and microwave) will not be considered in this text. Nuclear magnetic resonance spectroscopy involves the absorption of microwave radiation.

Absorption of Radiation

A qualitative understanding of the absorption of radiation can be obtained by considering the absorption of light in the visible region.

Table 12-1

The Electromagnetic Spectrum

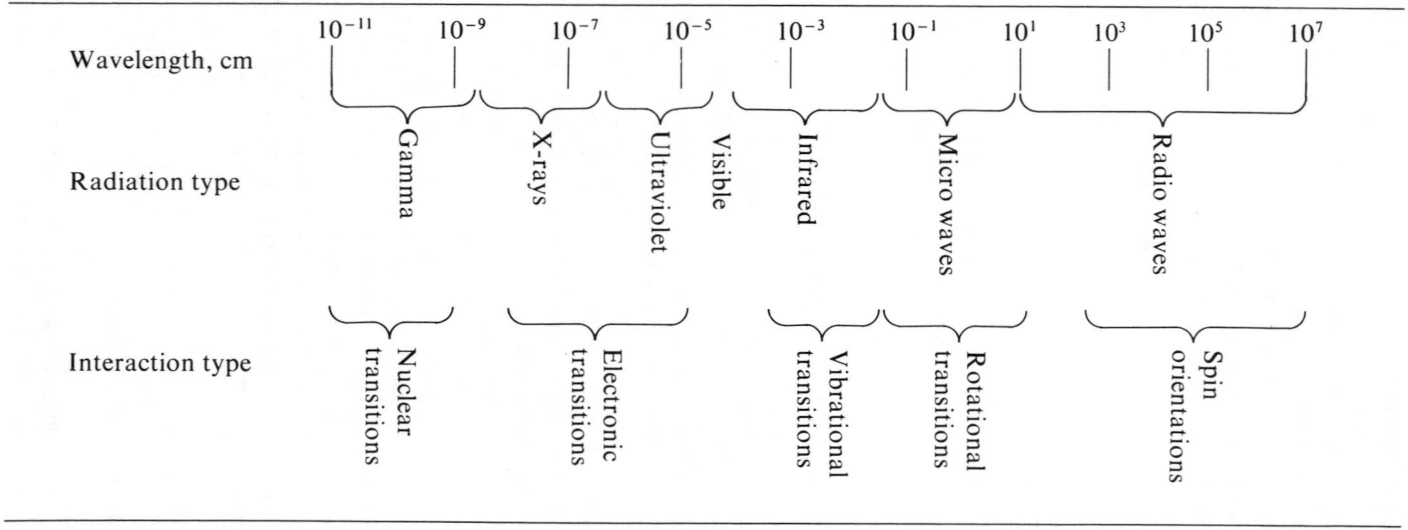

Wavelength, cm	Radiation type	Interaction type
10^7		
10^5	Radio waves	Spin orientations
10^3		
10^1	Micro waves	Rotational transitions
10^{-1}		
10^{-3}	Infrared	Vibrational transitions
10^{-5}	Visible	
	Ultraviolet	Electronic transitions
10^{-7}	X-rays	
10^{-9}	Gamma	Nuclear transitions
10^{-11}		

When white light, which contains the whole spectrum of wavelengths in the visible region, is passed through an object, certain wavelengths of light will be absorbed leaving unabsorbed wavelengths to be transmitted. The transmitted wavelengths are seen as a color which is complementary to the absorbed colors. Similarly, opaque objects absorb certain wavelengths leaving a residual color to be reflected and seen. Table 12-2 lists the approximate wavelengths associated with different colors in the visible spectrum.

Table 12-2

Colors of Wavelength Regions

Wavelength, nm	Absorbed color	Transmitted color
380–450	Violet	Yellow-green
450–495	Blue	Yellow
495–570	Green	Violet
570–590	Yellow	Blue
590–620	Orange	Green-blue
620–750	Red	Blue-green

There are three basic processes by which a molecule can absorb radiation. All three processes involve raising the internal energy level of the molecule equal to the energy of the absorbed radiation. The changes which occur in a molecule can be *electronic,* where the electrons are raised to a higher energy level, *vibrational,* which involves a change in the average separation of nuclei of two or more atoms, and *rotational,* which involves a change in the energy of the molecule as it rotates about its various axes.

Each of the three types of internal energy transitions are quantized, that is, they can occur only at definite wavelengths corresponding to an energy value ($h\nu$) equal to the jump in internal energy. There can be several different possible energy levels for each type of transition, and therefore, several different wavelengths may be absorbed. Rotational transitions occur at low energies (long wavelengths) in the far infrared and microwave region. At higher energy levels (shorter wavelength), in the mid infrared region, vibrational transitions occur. These vibrational transitions in combination with rotational transitions give rise to the infrared spectra of molecules. At still higher energy levels (visible and ultraviolet region) electronic transitions occur.

In a spectrophotometric analysis, a sample is irradiated with a beam

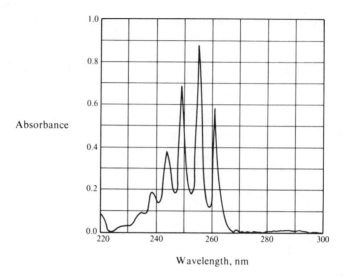

Figure 12-2

Ultraviolet absorption spectrum of benzene in cyclohexane.

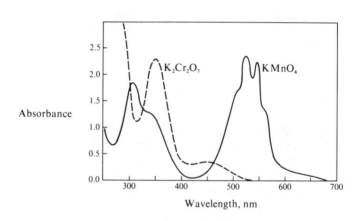

Figure 12-3

Visible absorption spectrum of $0.001M$ potassium
permanganate and $0.001M$ potassium dichromate in
$1M$ sulfuric acid.

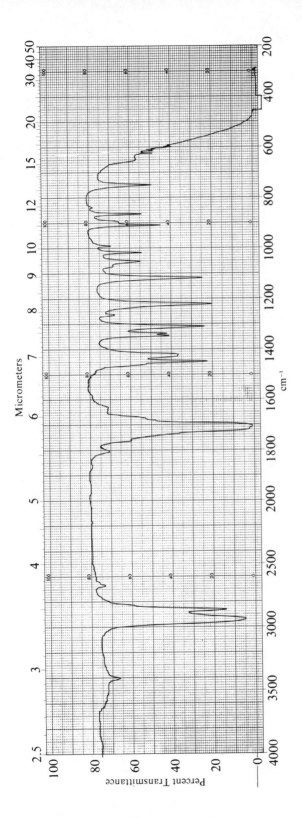

Figure 12-4

Infrared absorption spectrum of cyclohexanone.

of radiant energy of a narrow wavelength range and the amount of absorbed energy is measured. The wavelength of the incident beam is varied and the absorbance (or transmittance) is plotted against the wavelength. In such a manner an *absorption spectrum* is obtained. The instrument used to obtain the spectrum is called a spectrophotometer. Figures 12-2, 12-3, and 12-4 show typical spectra for compounds absorbing in the ultraviolet, visible and infrared regions of the spectrum.

Laws of Absorption

Quantitative methods of spectrophotometry are based upon the measurement of intensities or powers of radiation. The term power of radiation refers to the number of photons per unit time per unit volume and not the amount of energy per photon.

The power of transmitted radiation through a solution depends upon the power of the incident radiation striking the solution and the number of radiation absorbing particles which the radiation encounters. The number of particles encountered depends upon the thickness of solution through which the radiation must travel and the concentration of the radiation absorbing species. These factors are illustrated in Figure 12-5. Although it is experimentally difficult to measure the absolute power of radiant energy, it is relatively simple to measure the ratio between the incident power and the transmitted power. This ratio, P_o/P_t, is used to determine the concentration of a radiation absorbing species in a test solution.

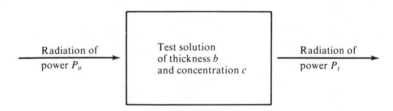

Figure 12-5

Absorption of radiation.

Lambert's Law or Bougher's Law

The law relating absorption of radiation to thickness of the absorbing medium was first formulated by Bougher in 1729. In 1760, Lambert restated the law and he is often given credit for its discovery. According to this law, each layer of absorbing medium of equal thickness absorbs an equal fraction of the radiation which travels through it. When a monochromatic beam of power P traverses the thickness of absorber b, the reduction in intensity is given by

$$-dP = k_1 P db \tag{12-1}$$

or

$$\frac{-dP}{db} = k_1 P \tag{12-2}$$

If we rearrange Equation (12-2) and integrate between the limits P_o and P_t (the initial and final radiant power of the light) and between zero and b for the width of absorbing solution we obtain

$$\int_{P_o}^{P_t} \frac{dP}{P} = -k_1 \int_o^b db \tag{12-3}$$

$$\ln \frac{P_t}{P_o} = -k_1 b \tag{12-4}$$

Equation (12-4) can also be written in the form

$$P_t = P_o e^{-k_1 b} \tag{12-5}$$

where e is the base of natural logarithms, b is the length of solution in cm through which the radiation traverses, and k_1 is a constant characteristic of the radiation absorbing species. Equations (12-4) and (12-5) are forms of Lambert's or Bougher's law.

Beer's Law

The law relating absorption of light to the concentration of the light absorbing species in a solution was formulated by Beer in 1859. Beer's law states that the power of a beam of monochromatic light decreases exponentially as the concentration of the absorbing species increases. The mathematical expression is

$$-dP = k_2 P dc \qquad (12\text{-}6)$$

Rearranging Equation (12-6) and integrating as before we obtain

$$\int_{P_o}^{P_t} \frac{dP}{P} = -k_2 \int_o^c dc \qquad (12\text{-}7)$$

$$\ln \frac{P_t}{P_o} = -k_2 c \qquad (12\text{-}8)$$

Equation (12-8) can also be written in the form

$$P_t = P_o e^{-k_2 c} \qquad (12\text{-}9)$$

where c is the concentration and all other terms are as defined previously. Equations (12-8) and (12-9) are forms of Beer's law.

Combined Law

Equations (12-4) and (12-8) or (12-5) and (12-9) can be combined into a single expression known variously as the Lambert-Beer law, the Bougher-Beer law, or simply Beer's law. The combined law is

$$P_t = P_o e^{-k_3 bc} \qquad (12\text{-}10)$$

or

$$\ln \frac{P_t}{P_o} = -k_3 bc \qquad (12\text{-}11)$$

Converting Equation (12-11) from natural logarithms to base 10 logarithms we obtain

$$\log \frac{P_t}{P_o} = \frac{-k_3 bc}{2.303} \qquad (12\text{-}12)$$

The term $k_3/2.303$ in Equation (12-12) is a constant and is given the symbol a when the concentration c is in grams per liter, or ϵ when c is in moles per liter. We therefore have

$$-\log \frac{P_t}{P_o} = abc \qquad (c, \text{g/l}) \qquad (12\text{-}13)$$

or

$$-\log \frac{P_t}{P_o} = \epsilon bc \qquad (c, \text{m/l}) \qquad (12\text{-}14)$$

The symbol a is called the *absorptivity* and ϵ is called the *molar absorptivity*.

The ratio P_t/P_o is known as the *transmittance* and is given the symbol T. We therefore have

$$-\log T = abc \qquad (c, \text{g/l}) \qquad (12\text{-}15)$$

or

$$-\log T = \epsilon bc \qquad (c, \text{m/l}) \qquad (12\text{-}16)$$

We can invert the log ratio in the above equations and remove the negative sign.

$$-\log \frac{P_t}{P_o} = \log \frac{P_o}{P_t}$$

The term $\log(P_o/P_t)$ is defined as *absorbance* and is given the symbol A. Thus,

$$A = abc \qquad (c, \text{g/l}) \qquad (12\text{-}17)$$

or

$$A = \epsilon bc \qquad (c, \text{m/l}) \qquad (12\text{-}18)$$

Many instruments are calibrated to read both absorbance and percent transmittance ($\% T$). Percent transmittance is the fraction transmitted times 100. Absorbance is more convenient to use because it has a linear relationship to concentration. Absorbance and percent transmittance vs. concentration at a given wavelength and cell pathlength are plotted in Figure 12-6.

Obviously absorbance and percent transmittance are related to each other. From Equations (12-16) and (12-18) we see that

$$A = -\log T$$

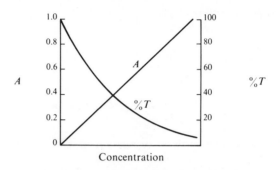

Figure 12-6

Absorbance and percent transmittance vs.
concentration at a given wavelength
and cell pathlength.

or

$$A = 2 - \log \% T \qquad (12\text{-}19)$$

Instruments which read both absorbance and percent transmittance use a linear scale for $\% T$ and a logarithmic scale for A. With this arrangement, a more precise reading can be made for A when the absorbance is low and a more precise reading for $\% T$ when the absorbance is high. Therefore, at absorbancies above approximately 0.6, it is preferable to read $\% T$ and calculate A by use of Equation (12-19).

EXAMPLE 12-1: A solution transmits 11.2% of the incident radiation at a given wavelength. Calculate the absorbance of the solution.

$$A = 2 - \log 11.2$$
$$= 2 - 1.049$$
$$= 0.951$$

EXAMPLE 12-2: A 0.0292 g sample of $MnCO_3$ was dissolved in acid, oxidized to the permanganate ion, and diluted to 1.000 liter. The absorbance of this solution at 535 nm was 0.526 in a 1.000 cm cell. Calculate the molar absorptivity of the permanganate ion at 535 nm.

The molarity of the permanganate solution is

$$M = \frac{0.0292 \text{ g/l}}{\text{MnCO}_3 \text{ g/m}} = 2.54 \times 10^{-4} M$$

$$A = \epsilon bc$$

$$\epsilon = \frac{0.526}{(1.000)(2.54 \times 10^{-4})}$$

$$= 2070$$

EXAMPLE 12-3: Compound X has a molar absorptivity of 8.25×10^3 at 560 nm. Calculate the concentration of X in a solution which transmits 22.0% of the radiation when placed in a 1.17 cm absorption cell.

$$A = 2 - \log 22.0$$

$$= 0.658$$

$$c = \frac{A}{\epsilon b}$$

$$= \frac{0.658}{(8.25 \times 10^3)(1.17)}$$

$$= 6.82 \times 10^{-5} M$$

Deviations from Beer's Law

When the plot of absorbance versus concentration for an absorbing species at a particular wavelength is linear, the substance is said to obey Beer's law (see Figure 12-7). Such a plot is known as a calibration curve or working curve. When a calibration curve is not linear the substance is said to deviate from Beer's law. True deviations from Beer's law occur only in systems where the concentration of the absorbing species is so high that the index of refraction for the absorbed radiation is changed. Apparent deviations from Beer's law may result from instrumentation limitations or effects of nonsymmetrical chemical equilibrium.

Instrumental variations which may cause apparent deviations to Beer's law include: power fluctuations of the radiation source and detector amplification system, sensitivity changes in the detector, and stray radiation reflected within the instrument reaching the detector. These errors are largely canceled out by use of a double beam spectro-

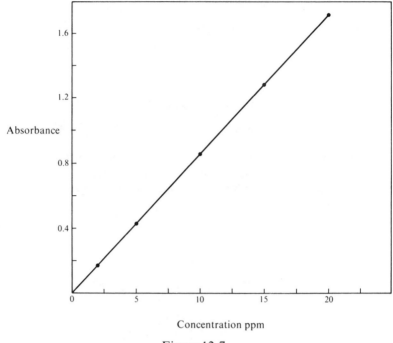

Figure 12-7

Calibration curve for aqueous potassium
permanganate solutions measured at 525 nm.

photometer but can occur using a single beam spectrophotometer. A
more serious instrumental source of deviation results from using a
narrow band of wavelengths rather than truly monochromatic (single
wavelength) radiation.

Nonsymmetrical chemical equilibria cause apparent deviations from
Beer's law because the relative proportions of the two absorbing species
change with concentration. These changes may result from inter-
molecular interactions, concentration-dependent dissociations or associations, reactions with solvent or hydrogen ion, or formations of
complex ions with varying number of ligands. Three examples are
given.

Copper(II) forms a series of complexes with ammonia having the
formulas $Cu(NH_3)(H_2O)_3^{++}$, $Cu(NH_3)_2(H_2O)_2^{++}$, $Cu(NH_3)_3(H_2O)^{++}$,
and $Cu(NH_3)_4^{++}$. Each complex has a different molar absorptivity at
the wavelength used for analysis. Unless a large excess of ammonia is
used the relative proportion of each species will be dependent upon the

total copper and total ammonia concentration. As a consequence in the spectrophotometric analysis of inorganic complexes the normal technique is to add a large excess of ligand. Also hydrogen ions compete with the metal ion for complexation with many ligands and it is often necessary to adjust the pH to a rather high value to ensure complete metal ion-ligand complexation.

Aqueous solutions of potassium chromate or potassium dichromate also show an apparent deviation from Beer's law because of interaction with hydrogen ions. The reaction is

$$2CrO_4^= + 2H^+ \rightleftharpoons 2HCrO_4^- \rightleftharpoons Cr_2O_7^= + H_2O$$

All three species ($CrO_4^=$, $HCrO_4^-$, and $Cr_2O_7^=$) absorb radiation and have different absorptivities. When unbuffered aqueous solutions of chromate or dichromate are diluted with water the *ratio* of chromate to dichromate will change causing an apparent deviation. If the solution is made strongly basic, the Cr(VI) is present essentially completely as $CrO_4^=$ and Beer's law is obeyed.

Weak acids which at a given wavelength absorb only in the undissociated form or only in the dissociated form or both forms to a different degree also may exhibit an apparent deviation from Beer's law. In an unbuffered solution, the pH may change with changes in concentration and the ratio of dissociated to undissociated forms would change causing a nonlinear Beer's law curve.

The above three examples illustrate the importance of understanding the chemical system being studied and controlling the conditions so that meaningful absorption measurements can be obtained.

Analysis of Mixtures

The spectrophotometric determination of two light absorbing species is easy if there is a wavelength where only one of the species absorbs and a second wavelength where only the second species absorbs. The absorbances at the two wavelengths are measured and the concentrations are calculated from the respective molar absorptivities. This situation is illustrated in Figure 12-8.

Spectrophotometric analysis of two absorbing species is also easy when only one species absorbs at one analytical wavelength and both species absorb at the second. This situation is shown in Figure 12-9. In this case the molar absorptivity of pure Y is measured on a known concentration of Y at both λ_1 and λ_2. The molar concentration of Y in

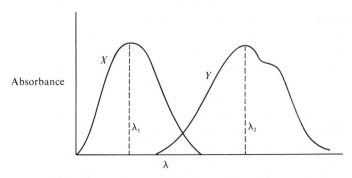

Figure 12-8

Spectrophotometric analysis of a mixture of X and Y,
no mutual absorbance at either analytical wavelength.

the mixture is determined from the absorbance at λ_2 and its ϵ at λ_2. The absorbance at λ_1 due to Y is calculated from its ϵ at λ_1 and its now known concentration. The remaining absorbance at λ_1 is due to X and its concentration is calculated from its molar absorptivity at λ_1. Obviously the experimental error in determining the concentration of X in this manner is greater than in the previous case where no mutual absorbance occurred.

Spectrophotometric analysis of two absorbing species, both of which absorb at both analytical wavelengths, requires the solution of two simultaneous equations in two unknowns. An example of this is shown in Figure 12-10. The total absorbance at each analytical wavelength

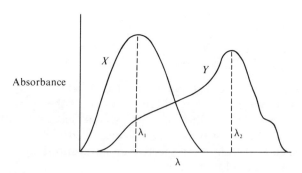

Figure 12-9

Spectrophotometric analysis of a mixture of X and Y,
both absorbing at one analytical wavelength, and only
one absorbing at the other.

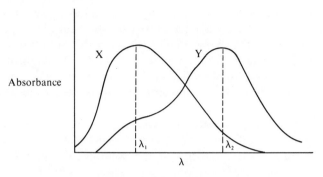

Figure 12-10

Spectrophotometric analysis of a mixture of X and Y,
both absorbing at each analytical wavelength.

is the sum of the absorbances of each absorbing component. The
equations are

$$A_{(\lambda 1)} = \epsilon_{X(\lambda 1)} bc_X + \epsilon_{Y(\lambda 1)} bc_Y$$

$$A_{(\lambda 2)} = \epsilon_{X(\lambda 2)} bc_X + \epsilon_{Y(\lambda 2)} bc_Y$$

Where: $A_{(\lambda 1)}$ = measured absorbance at λ_1
$A_{(\lambda 2)}$ = measured absorbance at λ_2
$\epsilon_{X(\lambda 1)}$ = molar absorptivity of X at λ_1
$\epsilon_{X(\lambda 2)}$ = molar absorptivity of X at λ_2
$\epsilon_{Y(\lambda 1)}$ = molar absorptivity of Y at λ_1
$\epsilon_{Y(\lambda 2)}$ = molar absorptivity of Y at λ_2
c_X = molar concentration of X
c_Y = molar concentration of Y
b = path length in cm

Solution of the two equations gives the molar concentration of X and
Y. The ϵ values must be obtained by measurements on pure solutions
of X and Y at the two wavelengths. Theoretically equations can be
set up for any number of components as long as absorbance readings
are taken at as many wavelengths. Small errors in measurements are
magnified by the calculation involved and the method is generally
limited to two or perhaps three component systems.

General Considerations on Spectrophotometry

Several questions need to be considered before a quantitative method
based on spectrophotometry can be used.

(1) Selection of wavelength. Generally absorbance measurements are
     made at a wavelength corresponding to an absorption peak. At
     this wavelength the change in absorbance with concentration is the
     greatest. Therefore, the maximum sensitivity is obtained. The
     absorption curve is often flat in this region; thus, the measure-
     ments will be less sensitive to errors due to nonexact reproduction
     of the wavelength setting of the instrument.
(2) Variables which affect the absorbance. A number of common
     variables often influence the absorption spectrum of a substance.
     Some examples are the pH of the solution, the nature of the
     solvent, the temperature, and the presence of interfering species.
     The effects of these variables must be known and controlled.
(3) Development and stability of color. The time required for colored
     species to reach full development must be determined. Most
     species form very rapidly but others need a certain development
     time to reach full color. The color of some species is stable for
     long periods of time but other species begin to fade after a while.
     This effect must be determined and absorbance readings made
     when the color is at a maximum.
(4) Adherence to Beer's law. Once the conditions for the method have
     been determined, it is necessary to prepare a calibration curve
     from a series of standard solutions. These standards should
     approximate the overall composition of the samples and should
     cover a range of concentrations on both sides of that of the species
     being determined.

Photometric Titrations

Measurements of absorbance can be employed to follow the progress
of many titrations. Photometric titrations may be performed whenever
(a) the substance being titrated, (b) the product formed in the titration,
or (c) the titrant itself absorb strongly at a given wavelength. The
titration curve involves a plot of absorbance as a function of titrant
volume and consists of two straight lines. The intersection of these two
lines is the end point. Typical photometric titration curves are shown
in Figure 12-11. Often data points near the equivalence point fail to
fall perfectly on the two straight-line portions of the titration curve.
This is due to the fact that the titration may be appreciably incomplete
at the equivalence point. This behavior is indicated by the dotted line
portion of curve *A* in Figure 12-11. To avoid a possible error due to
this effect enough points are taken sufficiently on each side of the end

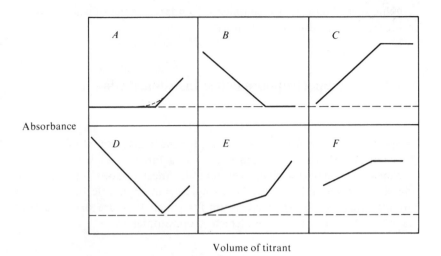

Figure 12-11

Photometric titration curves.

point to determine the straight lines and they are extrapolated to determine the end point. In Figure 12-11, the curves represent the following: Curve A is the titration of colorless substance with a colored titrant yielding a colorless product; curve B, titration of a colored substance with a colorless titrant yielding a colorless product; curve C, titration of a colorless substance with a colorless titrant yielding a colored product; curve D, titration of a colored substance with a colored titrant yielding a colorless product; curve E, titration of a colorless substance with a colored titrant yielding a colored product; and curve F, titration of a colored substance with a colorless titrant yielding a product with a higher ϵ than the substance titrated.

If in the photometric titration plot, the volume of titrant added is appreciable compared to the volume of the solution, it is necessary to correct measured absorbance values for the dilution effect. If the

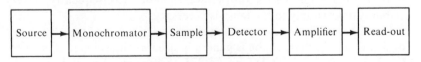

Figure 12-12

Block diagram of essential components of a
single beam spectrophotometer.

volume spread plotted is insignificant compared to the solution volume, no correction for dilution is required.

Spectrophotometric Instrumentation

A spectrophotometer is an instrument for the measurement of transmittance or absorbance of radiant energy as a function of wavelength or for measurements at a given wavelength. Spectrophotometers may be classified as manual or recording, single beam or double beam. They are also classified by spectral region for which they are designed such as infrared, visible, or ultraviolet spectrophotometers.

Single Beam Spectrophotometers

The essential components of a manually operated, single beam spectrophotometer are shown in a block diagram in Figure 12-12. All spectrophotometers require (1) a *source* of continuous radiation over the wavelengths of interest; (2) a monochromator; (3) a sample container; (4) a detector; (5) an amplifier; and (6) a read-out device.

Sources

The usual source for radiant energy for visible spectrophotometry as well as the near infrared is an incandescent lamp with a tungsten filament. The useful wavelength range is from about 325 or 350 nm to about 3 μm. The radiant energy emitted by the heated filament varies with wavelength as shown in Figure 12-13. The energy distribution is a function of the temperature of the filament which in turn depends upon the voltage supplied to the lamp. An increase in voltage increases the temperature, which increases the total energy output and shifts the peak of Figure 12-13 to a shorter wavelength. Consequently, a very stable voltage supply to the lamp is necessary.

For the ultraviolet region of the spectrum, a low pressure hydrogen or deuterium discharge tube is generally used as the source. The effective range is from about 185 nm to 375 or 400 nm. The hydrogen discharge tube consists of a pair of electrodes in a glass tube containing hydrogen gas at low pressure. A high applied voltage causes an electron discharge which excites other electrons to higher energy levels. As the electrons return to their ground state they emit continuous radia-

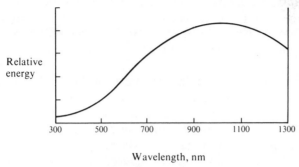

Wavelength, nm

Figure 12-13

Relative energy output of a tungsten lamp
as a function of wavelength.

tion in the ultraviolet region. A quartz window is provided in the glass tube to allow transmittance of ultraviolet radiation. Many spectrophotometers are provided with a mechanism for interchanging tungsten and hydrogen discharge tubes in order to make measurements over both the visible and ultraviolet regions.

In the infrared region of the spectrum, most prism spectrophotometers operate from about 2 μm to 15 μm and grating spectrophotometers operate at 2.5 μm to 25 μm. The usual radiation source is the Nernst glower or the Globar. The Nernst glower is a hollow rod of zirconium and yttrium oxides which when heated to 1500°C by an electric current emits radiation in the range 0.4 to 20 μm. The Globar is a silicon carbide rod which when heated to 1200°C emits radiation in the 1–40 μm range. The Globar is a more stable source than the Nernst glower.

Monochromators

As indicated above, radiation sources employed in spectrophotometry emit continuous radiation over a wide range. A narrow band width of radiation is required (1) to allow resolution of absorptions bands which are close to each other, (2) to allow the measurement of an absorption maximum at its peak, thereby increasing the sensitivity, and (3) so that Beer's law will be more closely obeyed.

Two types of devices are employed to resolve wide band polychromatic radiation into narrow bands. These devices are known as filters and monochromators. A filter allows transmission of limited wavelength regions while absorbing most of the radiation of other wavelengths. Filters typically transmit radiation with an effective band width (where the transmittance is at least one-half its maximum value) of from 20 to 50 nm. A monochromator resolves polychromatic radiation into an effective band width of from a few tenths of a nanometer to about 20 nm.

The components of a monochromator include: (1) an entrance slit which admits polychromatic light from the source; (2) a collimating lens or mirror; (3) a dispersion device, either a prism or grating, which resolves the radiation into its component wavelengths; (4) a focusing lens or mirror; and (5) an exit slit. All the components must be transparent in the wavelength range being used. A schematic diagram of a prism monochromator is shown in Figure 12-14. The spectral purity of the emergent radiation depends on the dispersive power of the prism or diffraction grating and the widths of the entrance and exit slits. The narrower the slit width, the narrower the wavelength band. However, narrowing the slit width also lowers the amount of radiation which strikes the detector and a physical limit is imposed upon the slit width due to the sensitivity of the detector employed.

Sample containers

The cell holding the sample must be transparent in the wavelength region being used. Ordinary glass may be used for visible measurements but quartz or fused silica cells must be used in the ultraviolet region because glass absorbs ultraviolet radiation.

Samples studied in the ultraviolet or visible region are normally solutions. The typical path length for sample cells is 1 to 10 cm. Microcells with a path length of 0.1 cm are available for small samples which cannot be diluted. The windows of the absorption cells must be kept scrupulously clean; fingerprint smudges and traces of contamination can cause considerable error. Small scratches can cause scattering of the light beam with a corresponding error. The sample cell, when positioned, becomes part of the optical path through the spectrophotom-

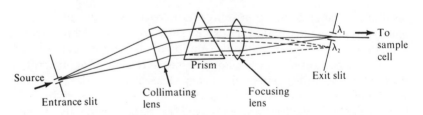

Figure 12-14

A prism monochromator.

eter and must be positioned exactly the same each time. The cell must be filled such that the radiation passes through the solution and not the meniscus. The solution should be free of air bubbles.

For infrared spectrophotometry, samples may be gases, liquids, or solids. Gas samples are held in glass tubes having sodium chloride, potassium bromide, or calcium fluoride windows all of which are transparent in the region studied. Path lengths for gaseous samples vary from a few centimeters to several meters obtained by multiple reflections in the cell. Pure liquids are run as a thin film (0.01–0.05 mm) between NaCl, KBr, or CaF$_2$ salt plates. Solutions are run in salt plate cells with a path length of 0.1 to 1 mm. Chloroform, carbon tetrachloride, and carbon disulfide are the common solvents used in infrared spectrophotometry. Solids are either mixed with KBr and pressed into a thin disc or run as a suspension known as a mull in a viscous liquid such as Nujol, a mineral oil.

Detectors

A detector absorbs the energy of the photons which strike it and converts this energy into electrical energy. For the visible or ultraviolet region a *phototube* or *photomultiplier tube* is employed. In the phototube, photons cause electrons to be emitted from the cathode. The electrons are attracted to the anode which causes a current to flow. A photomultiplier tube is essentially a series of phototubes built into one. The current produced by the photons is amplified several fold in a photomultiplier tube. The common detector used for the infrared region is a *thermocouple* where the heat of the infrared radiation causes an electrical signal to be generated.

Amplification and Readout

The detailed electronics of amplification and readout are beyond the scope of this text. Let it suffice to say that the process is accomplished with amplifiers, ammeters, potentiometers, and potentiometric recorders.

Operation of a Single Beam Spectrophotometer

To standardize a single beam spectrophotometer the instrument must be adjusted such that the meter reads 0% transmittance or infinite absorbance when no radiation reaches the detector and 100% transmittance or zero absorbance when the radiation passes through a cell containing none of the radiation absorbing species. An opaque shutter, controlled by the operator, is placed in front of the phototube to prevent radiation from reaching it. Even with the phototube in complete darkness a small current, known as the *dark current,* flows in the phototube due to random thermal emission of electrons from the cathode surface. A knob is adjusted to cancel out the dark current, setting the meter to read infinite absorbance or $0\% T$. Next, with the wavelength set at a desired value, a cell containing a reference solution is placed in the radiation beam. The reference solution is generally the pure solvent or a "blank" which contains all the reagents used in the procedure but no sample. The shutter is removed to allow the radiation to reach the detector. The instrument is now adjusted to read $100\% T$ or zero absorbance. This adjustment is made by changing the monochromator slit control and/or the gain of the amplifier. Now the reference solution is replaced by the sample solution and the percent transmittance or absorbance is read from the meter.

When the wavelength is changed the reference solution setting must be again adjusted to $100\% T$ or zero absorbance to account for the variation of source output or detector response, both of which are wavelength dependent. The dark current should be checked periodically in case of possible drift in the circuit. Normally two cells are used, one for the reference, the other for the sample. These cells should be matched for path length and optical quality. This match can easily be checked by filling each with an identical solution (usually solvent) and checking one against the other throughout the wavelength region being used.

Double Beam Spectrophotometers

A double beam spectrophotometer can be used to automatically plot the absorbance of a sample as a function of wavelength. The details of instrumentation is not covered in this text but a simple explanation of the theory follows.

Radiation from the source passes through a monochromator and strikes a chopper, which is a rotating mirror device that permits half the radiation to pass through and the other half to be reflected off at a right angle. Through a system of mirrors, the one beam passes through a cell containing the reference solution and the other beam passes through a cell containing the sample. The two beams are reunited by mirrors so that they strike the detector. The reference and sample beams strike the detector alternately and the instrument records the ratio of the two signals.

Differential Spectrophotometry

A method is available for increasing the accuracy of measuring absorbancies. This method is known as differential spectrophotometry. In conventional spectrophotometry the instrument is set to read $100\% T$ or zero absorbance with a blank solution or solvent as reference. Actually any solution can be placed in the radiation beam and the instrument adjusted to $100\% T$. If we use a standard solution of the radiation absorbing species, which is less concentrated than the unknown, as the reference, we can adjust the instrument to read $100\% T$ for the standard. Then, the $\% T$ of the unknown is measured compared to the standard. This results in a scale expansion as shown in Figure 12-15 with an enhanced accuracy in the reading.

In the example shown in Figure 12-15, a standard solution which has a 25% transmittance using solvent as a reference is used as the reference. An unknown solution measured against the standard solution has a 72% transmittance. This means the solution transmits 72% as much radiation as the standard. If the unknown were measured using solvent as a reference it would read $18\% T$. The relative error in measurement employing differential spectrophotometry in this case would be four times smaller. Another application of differential

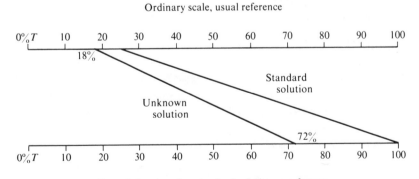

Figure 12-15

Scale expansion for differential spectrophotometry.

spectrophotometry is in the measurement of solutions which are too highly absorbing for conventional measurements.

An ultimate method exists for increasing the accuracy of a spectro-photometric measurement. This involves setting the $100\% T$ with a solution more dilute than the unknown as described above and setting the $0\% T$ with a more concentrated standard solution in the beam rather than the shutter. The unknown is then compared to the two standard solutions. For further details Reilley and Crawford should be consulted.*

Questions

1. What is meant by the dual nature of light?

2. Define the following terms: frequency, wave number, nanometer, mono-chromatic light, molar absorptivity, Beer's law, photometric titration.

3. Name the three basic processes by which molecules absorb radiation and tell which region of the electromagnetic spectrum these absorptions occur.

4. If a solution absorbs blue light, what color will it appear to the eye?

5. Discuss why some species fail to follow Beer's law.

6. Why are most spectrophotometric measurements made at absorption maxima?

7. Name the essential components of a manual spectrophotometer and explain the function of each.

8. What two devices can be used as a monochromator?

9. Why must glass absorption cells not be used for ultraviolet measurements?

10. How can the spectral purity of a light beam be improved with an existing spectrophotometer, and what problem does this create?

Problems

1. Calculate the frequency in reciprocal centimeters and the energy in ergs of a photon of yellow light having a wavelength of 475 nm.

2. Calculate the absorbance of solutions which have the following values of percent transmittance: (a) 75, (b) 52, (c) 10, (d) 4.1, (e) 1.0.

3. Convert the following absorbance readings to percent transmittance: (a) 0.005, (b) 0.23, (c) 0.48, (d) 1.00, (e) 1.64, (f) 2.0.

4. A solution was found to have a 15.6% transmittance at its wavelength of

---

*C.N. Reilley and C.M. Crawford, *Anal. Chem.*, **27**, 716 (1955).

maximum absorption using a cell with a path length of 5.000 cm. Calculate the percent transmittance and absorbance of the solution in (a) a 10.0 cm cell, (b) a 1.0 cm cell.

5. A compound has a molar absorptivity of 9,300 at 395 nm. Calculate the molarity of a solution of this compound which has an absorbance of 0.425 when measured in a 1.000 cm cell.

6. A solution contains 4.0 parts per million iron. The iron is reduced to the ferrous state and is complexed with 1,10-phenanthroline. The absorbance of the solution measured at 508 nm in a 1.000 cm cell is 0.795. Calculate the molar absorptivity of the ferrous-1,10-phenanthroline complex at this wavelength.

7. A 1.000 g sample of steel is dissolved in nitric acid and the manganese is oxidized to permanganate with potassium periodate. The resulting solution is diluted to 500.0 ml and has an absorbance of 0.62 at 530 nm using a 1.000 cm cell. A manganese standard was prepared by dissolving 0.0330 g of $MnCO_3$ in acid and oxidizing the manganese to permanganate. The standard solution was diluted to 1000.0 ml and has an absorbance of 0.58. Calculate the percent manganese in the steel sample.

8. A compound having a molecular weight of 150 has a molar absorptivity of 15,300. Calculate the weight of the compound required to dissolve in 1.000 liter such that a 5.0 ml aliquot diluted to 100.0 ml will give an absorbance of 0.50 in a 1.000 cm cell.

9. The following data were obtained for a complex of nickel at 575 nm in a 1.000 cm cell:

Standard no.	Nickel, ppm	Absorbance
1	1.0	0.058
2	2.0	0.109
3	4.0	0.220
4	6.0	0.328
5	10.0	0.552

(a) Plot a calibration curve and determine the molar absorptivity of the complex.
(b) Calculate the concentration of an unknown solution having an absorbance of 0.262.

10. The percent transmittance of a solution was recorded as 42.4 and the sample discarded. It was then discovered that the transmittance with the blank had been set at 95% transmittance rather than 100%. Calculate the correct transmittance for the solution.

11. Plot the absorption spectra for the violet colored complex of iron(III) with salicylic acid. Determine the best wavelength for a measurement of this complex.

Wavelength, nm	A	Wavelength, nm	A
350	0.398	520	0.708
370	0.287	530	0.712
390	0.215	540	0.703
410	0.220	550	0.684
430	0.297	570	0.627
450	0.416	590	0.552
470	0.535	610	0.460
490	0.640	630	0.374
510	0.701	650	0.320

12. A given acid-base indicator exists totally undissociated at pH 4.0 and totally dissociated at pH 11.0. A $4.25 \times 10^{-4}M$ solution of the indicator has an absorbance of 0.042 at pH 4.0 and an absorbance of 0.92 at pH 11.0. The same concentration of indicator has an absorbance of 0.484 at pH 7.50. All absorbance readings were measured at 460 nm. Calculate the concentration of the acidic and basic forms at pH 7.50.

13. Calculate the ionization constant for the weak acid indicator in the previous problem.

14. Compound A has a molar absorptivity of 12,700 at 410 nm and 4,200 at 575 nm. Compound B does not absorb at 410 nm and has a molar absorptivity of 7,280 at 575 nm. An unknown solution containing A and B has an absorbance of 0.650 at 410 nm and 0.580 at 575 nm using 1.000 cm cells. Calculate the concentration of A and B in the unknown solution.

15. A solution containing dichromate and permanganate ions in sulfuric acid was analyzed spectrophotometrically at 435 nm and 535 nm using 1.000 cm cells. From the following data calculate the dichromate and permanganate concentration in the unknown.

	Absorbance	
Solution	435 nm	535 nm
$2.0 \times 10^{-4}M$ $KMnO_4$	0.009	0.444
$1.2 \times 10^{-3}M$ $K_2Cr_2O_7$	0.470	0.015
unknown	0.463	0.268

16. Compound A has a molar absorptivity of 840 at 420 nm and 1820 at 545 nm. Compound B has a molar absorptivity of 2810 at 420 nm and 240 at 545 nm. An unknown solution containing A and B was analyzed in a 1.000 cm cell with the percent transmittance readings being 52.4% at 420 nm and 61.0% at 545 nm. Calculate the concentration of A and B in the unknown.

17. Iron in a sample can be determined by the photometric titration of the iron(III)-salicylate complex with EDTA. A 10.00 ml aliquot of a sample containing iron was converted into the complex, diluted to about 50 ml,

and titrated with $0.01500M$ EDTA. Plot the following data and determine the concentration of iron in parts per million in the sample.

ml EDTA	Absorbance
36.50	0.684
37.00	0.580
37.50	0.481
38.00	0.378
38.50	0.270
39.00	0.166
39.50	0.060
40.00	0.013
40.50	0.013
41.00	0.013

18. The following data were obtained for the photometric titration of 5.00 ml of a colorless metal ion with $0.0120M$ solution of a colorless complexing agent. The reaction which occurs results in the formation of a complex which absorbs light at 580 nm. No water was added during titration and a 10 ml buret was used for accuracy. The absorbance readings were as follows:

ml complexing agent	Absorbance @ 580 nm
0.00	0.000
1.00	0.064
2.00	0.110
3.00	0.142
4.00	0.168
5.00	0.163
6.00	0.148
7.00	0.135

Plot the data (see problem 19) and calculate the millimoles of metal ion in the sample.

19. Comment on why no correction for dilution was required in the plotting of data for problem 17 but was required for the *correct* plotting of problem 18.

20. A $1.24 \times 10^{-3}M$ solution of a light absorbing species gives 28.2% transmittance using water as a blank. The spectrophotometer is adjusted to read 100% transmittance with the above solution as a blank and an unknown solution read against this blank has a 38.2% transmittance.
    (a) Calculate what the percent transmittance of the unknown solution would be using water as a blank.
    (b) Calculate the concentration of light absorbing species in the unknown.

Determination of Water in a Hydrate

Principle: The determination of chemically bound water in a hydrate is an analysis in which water is determined indirectly by the loss in weight upon drying the sample. This method can be extended to the determination of any volatile constituent in a sample. If two or more volatile constituents are present a direct method must be utilized to determine their respective percentages. In this experiment the sample will be dried to constant weight at 110°C.

PROCEDURE

1. Clean two glass stoppered weighing bottles. Number each weighing bottle and cover on the ground glass surface with a pencil. Dry the weighing bottle and cover for 30 minutes in the 110° oven. Cool the weighing bottles at least 30 minutes in the desiccator.
2. Weigh each weighing bottle and cover to the nearest 0.0001 g. Add

approximately 1 g of sample to each weighing bottle and again weigh to the nearest 0.0001 g.

3. Place the weighing bottles, covers, and samples in the 110° oven with the cover removed until the following lab period.

4. Transfer the weighing bottles to the desiccator and cool for 30 minutes.

5. Weigh the weighing bottles with their covers firmly in place to the nearest 0.0001 g.

6. Calculate the percent of water in each sample. Duplicate results should agree to within 0.10%. Report the average percent to the nearest 0.01%.

Experiment **2**

Gravimetric Determination of Chloride

Principle: The chloride in a soluble sample is precipitated with an excess of silver nitrate. After coagulation has occurred, the precipitate is transferred to a filter crucible, washed, dried, and weighed.

PROCEDURE

1. Clean and dry to constant weight three sintered glass filtering crucibles. Cool the crucibles for 30 minutes in a desiccator and weigh to the nearest 0.0001 g.
2. Dry the chloride sample for at least two hours at 110°C. Cool the sample for 30 minutes in a desiccator and weigh triplicate samples of 0.4 to 0.6 g (to the nearest 0.0001 g) into three 400 ml beakers.
3. Dissolve the sample in 100 ml of distilled water and add 1 ml of concentrated nitric acid. Use a separate stirring rod for each

311

beaker and leave it in the beaker throughout the procedure. Keep the solution protected with a watch glass as much as possible.

4. Assume the sample is pure sodium chloride and calculate the volume of $0.20 M$ silver nitrate required to precipitate the chloride.
5. Heat the solution to about 80°C and add the calculated amount of silver nitrate slowly with a Mohr pipet while stirring the solution. Then add an excess of 10%.
6. Keep the solution near boiling (do not allow to boil) for 30 minutes in order to coagulate the precipitate. As soon as coagulation has begun check the sample for complete precipitation by adding a few drops of silver nitrate. If a precipitate forms add 10% more silver nitrate, stir and check for completeness of precipitation again when coagulation has occurred.
7. Allow the solution to cool for approximately 30 minutes before filtering.
8. Prepare 500 ml of dilute nitric acid wash solution by adding 2 ml of concentrated HNO_3 to 500 ml of distilled water. Place the wash solution in a wash bottle and clearly label "HNO_3 wash solution."
9. Assemble a suction flask and trap as illustrated in Figure 2-13.
10. Place a weighed and numbered filter crucible in the suction apparatus and apply gentle suction. Decant the clear supernatant liquid through the crucible using a glass stirring rod. Do not allow the crucible to become more than two-thirds full.
11. Add about 10 to 15 ml of wash solution to the precipitate while in the beaker, stir well, allow the precipitate to settle, and decant the liquid through the crucible. Repeat this operation two or three times.
12. Hold the stirring rod across the beaker and direct a stream of wash solution at the precipitate to wash it into the crucible.
13. Remove the last traces of precipitate from the beaker and stirring rod with a rubber policeman.
14. Wash the precipitate in the crucible with several 5 ml portions of wash solution. Catch some of the last washing in a small tube suspended under the filter and add a drop of dilute HCl. If a precipitate forms, further washing is required.
15. Drain the crucible completely, place in a marked beaker, cover with a watch glass and hooks and dry at 110°C for two hours. Cool in a desiccator for at least 30 minutes and weigh. Repeat the heating, cooling, and weighing until constant weight within ± 0.2 mg is attained. An alternate procedure for heating to constant weight is to dry at 110°C for 24 hours or longer.

16.   Calculate the percent chloride to the nearest 0.01%.   Report the mean and standard deviation of the results.

Note.   Keep the silver chloride precipitate out of direct sunlight as much as possible to prevent photodecomposition from occurring.

Gravimetric Determination of Sulfate

Principle: The sulfate in a soluble sample is precipitated as barium sulfate from an acid solution. The precipitate is washed, filtered, dried, and weighed.

PROCEDURE

1. Clean and dry three numbered porcelain crucibles. Heat each crucible to about 600°C or higher (dull redness) for one hour. Allow the crucibles to cool slightly and transfer with crucible tongs to a desiccator. Weigh each crucible without its cover after a 30 minute cooling period. Reheat the crucibles for 30 minutes as before, cool, and reweigh. If the weights agree within 0.2 mg, they are assumed to be at constant weight and are ready for use.

2. Dry the sample in a weighing bottle for at least two hours at 110°C. Cool the sample for 30 minutes in a desiccator and weigh triplicate samples of 0.4 to 0.6 g (to the nearest 0.0001 g) into three 400 ml beakers.

3. Dissolve each sample in about 200 ml of distilled water, add 2 ml concentrated hydrochloric acid, and heat to near boiling.

4. Dissolve 3.9 grams of barium chloride dihydrate in 300 ml of distilled water and heat to near boiling (Note 1).

5. Quickly, but carefully, pour 100 ml of the hot barium chloride solution into the hot sample solution. Stir vigorously for about two minutes. Digest the precipitate for one to two hours at which time the supernatant solution should be perfectly clear and the precipitate settled to the bottom. Check for completeness of precipitation by adding a few drops of barium chloride solution.

6. Set up a filter assembly as discussed in Chapter 2. Use ashless filter paper.

7. Decant the supernatant liquid through the filter. Wash the precipitate in the beaker three or four times with small portions of hot distilled water. Transfer the precipitate, with a stirring rod and hot water, to the filter. Remove the last traces of precipitate from the beaker and stirring rod with a rubber policeman, transferring the precipitate into the filter.

8. Wash the precipitate on the filter with hot water until a few milliliters of filtrate give a negative test for chloride when tested with silver nitrate solution.

9. Carefully fold the filter paper around the precipitate and place in a weighed crucible.

10. Place the crucible on a clay triangle with the crucible cover slightly displaced to allow steam and gases to escape and air to enter. Slowly heat the crucible to dry the precipitate and paper. When the moisture is gone, increase the heat to char the paper without allowing the paper to burst into flames (Note 2).

11. When only a small ash remains from the paper, remove the cover completely, tilt the crucible to allow free access of air, and heat to a dull redness to remove any carbon residue (Note 3).

12. Allow the crucible to cool slightly in air, then 30 minutes in a desiccator. Weigh the crucible, repeat the heating and cooling, and again weigh the crucible. If the two weights do not agree within 0.2 mg, additional heating periods must be made until agreement is obtained.

13. Calculate the percent sulfate (or SO_3 if desired) to the nearest 0.01%. Report the mean and standard deviation of the results.

Note 1.   The formation and growth of large barium sulfate crystals is aided if the barium chloride solution is aged for at least 24 hours prior to its use.

Note 2. If the paper does burst into flames, immediately cover the crucible completely to extinguish the flame.

Note 3. Sometimes carbon from the filter paper will reduce some of the barium sulfate to barium sulfide. If this is suspected, cool the precipitate nearly to room temperature and add two or three drops of concentrated sulfuric acid. Ignite the precipitate strongly for 10 to 15 minutes to expel the excess sulfuric acid.

$$BaS + H_2SO_4 \rightarrow BaSO_4 + H_2S \uparrow$$

Experiment **4**

Gravimetric Determination of Iron

Principle: Iron is determined by precipitation as hydrous ferric oxide, $Fe_2O_3 \cdot xH_2O$, followed by filtration and ignition to ferric oxide, Fe_2O_3. The iron is oxidized to Fe(III) by hydrogen peroxide, bromine, or nitric acid prior to precipitation.

PROCEDURE

1. Prepare three porcelain crucibles as in Experiment 3.
2. Unless otherwise directed, do not dry the sample. Weigh three appropriate sized samples (consult instructor) into 400 ml beakers. Dissolve these in about 50 ml of distilled water to which has been added 5 ml of concentrated hydrochloric acid.
3. Heat the solutions to near boiling, add about 30 drops of concentrated nitric acid. The solution may turn dark due to the formation of the complex $FeSO_4 \cdot NO$. Continue heating until the darkened solution turns yellow and the oxides of nitrogen have been expelled.

317

4. Dilute the solution to about 200 ml with distilled water, heat nearly to boiling, and add slowly, with stirring, freshly filtered $6M$ ammonia until precipitation is complete (Notes 1, 2, and 3).

5. Digest for a few minutes; check for completeness of precipitation by adding a few drops of $6M$ ammonia to the supernatant liquid.

6. Decant the supernatant liquid through a coarse ashless filter paper (Whatman No. 41 or equivalent). Wash the precipitate with two 30 ml portions of hot 1% ammonium nitrate solution. Do not transfer much of the precipitate during this step.

7. Return the filter paper to the beaker in which the precipitate was formed. Add about 5 ml of concentrated hydrochloric acid to each and macerate the paper completely with a stirring rod (Note 4).

8. Dilute the solution to about 200 ml with distilled water and re-precipitate as in step 4.

9. Decant the filtrate through ashless filter paper and wash the precipitate in the beaker with several portions of hot 1% ammonium nitrate until the filtrate shows no more than a trace of chloride ion when tested with silver nitrate (Note 5).

10. Transfer the precipitate quantitatively to the filter. Final traces of precipitate can be removed from the beaker walls by scrubbing with a small piece of ashless filter paper.

11. Allow the precipitate to drain thoroughly (preferably overnight). Char the filter paper and ignite to constant weight as in Experiment 3.

12. Calculate the percent Fe, Fe_2O_3, or whatever is desired to the nearest 0.01%. Report the mean and standard deviation of the results.

Note 1. Upon standing, aqueous ammonia solutions attack glass containers and become contaminated with silica. Therefore, the $6M$ ammonia should be filtered and used the same day or stored in a plastic container.

Note 2. The ammonia should be added until its odor is unmistakably present above the solution.

Note 3. The precipitate should be reddish brown. If it is greenish-black, it indicates some Fe(II) remains and the sample must be redissolved in hydrochloric acid and again oxidized with nitric acid.

Note 4. The macerated filter paper keeps the gelatinous precipitate more separated, making the washing and filtration more complete and rapid. The basic solution must not be allowed to set overnight. If the final filtration cannot be completed the sample must be left overnight at this point.

Note 5. Before testing with silver nitrate acidify the filtrate with dilute nitric acid.

Standardization of 0.1 N NaOH

Principle: A carbonate free solution of sodium hydroxide is prepared and standardized against primary standard potassium acid phthalate.

PROCEDURE

1. Pipet 6–7 ml of clear, 50% solution of sodium hydroxide into 1 liter of CO_2-free (freshly boiled and cooled) distilled water. Mix thoroughly and store in a polyethylene bottle.
2. Dry 4–5 g of primary standard potassium acid phthalate in a weighing bottle for two hours (only) at 110°. Cool in a desiccator for at least 30 minutes.
3. Accurately weigh three 0.8–0.9 g samples into 250 ml flasks.
4. Dissolve the sample in approximately 75 ml of distilled water, add three drops of phenolphthalein indicator, and titrate to the first faint pink end point which persists for 20 seconds. Be sure to

wash down the sides of the flask near the end point and to split drops at the end point.

5. Calculate the normality of the NaOH. The three standardizations should have a range no greater than 0.0002 or 0.0003 N.

Determination of Total Acidity

Principle: The acidity of an impure potassium acid phthalate sample is determined by titration with standardized sodium hydroxide.

PROCEDURE

1. Dry the sample for two hours (only) at 110°. Cool for at least 30 minutes in a desiccator.
2. Weigh one 1.2 g sample into a 250 ml flask.
3. Dissolve the sample in about 75 ml of distilled water and titrate with standardized sodium hydroxide exactly as in the standardization of the sodium hydroxide.
4. Weigh two additional samples into 250 ml flasks adjusting the weight so as to require approximately 40 ml of titrant.
5. Titrate as before.
6. Calculate the percent potassium acid phthalate (KHP) in the sample to the nearest 0.01%.

Standardization of 0.1 *N* Hydrochloric Acid

Principle: Procedure A is a method of preparing a primary standard solution of hydrochloric acid by titration against pure sodium carbonate. Procedure B is a method of preparing a secondary standard solution by comparison with standard sodium hydroxide solution.

PROCEDURE A

1. Calculate the volume of concentrated hydrochloric acid (12*N*) needed to prepare the required volume of approximately 0.1 *N* solution. Pipet this volume into a storage bottle, dilute to volume, and mix thoroughly by shaking.
2. Dry 1 to 2 g of primary standard sodium carbonate at 110°C for at least two hours in a weighing bottle. Cool for 30 minutes in a desiccator.
3. Weigh three 0.2 to 0.24 g samples of the sodium carbonate into 250 ml flasks and dissolve in about 100 ml distilled water.

4. Add three drops of modified methyl orange indicator (Note 1) to each flask and titrate with the hydrochloric acid solution until the indicator changes from green to gray (Note 2).
5. Calculate the normality of the acid using an equivalent weight of 53.00 for sodium carbonate. Replicate standardizations should agree with 0.0002 N.

PROCEDURE B

1. Prepare the hydrochloric acid solution as in A-1.
2. Pipet 25.0 ml aliquots of the acid into three 250 ml flasks (Note 3).
3. Add three drops of phenolphthalein indicator and titrate with standardized sodium hydroxide solution to the first faint pink end point which persists for 20 seconds.
4. Calculate the normality of the hydrochloric acid solution.

Note 1. Modified methyl orange indicator is a mixed indicator containing a blue dye, xylene cyanole FF. The mixed indicator is green on the alkaline side and purple on the acidic side. The pH range over which the gray color is visible is small and occurs at the appropriate pH for the titration.

Note 2. Methyl red indicator can be used, but the solution must be boiled shortly before the end point to remove dissolved carbon dioxide.

Note 3. Samples of 40.0 ml can be used if a second buret is set up and the acid solution measured from the buret into the flasks.

Determination of Total Basicity of Impure Sodium Carbonate

Principle: Impure sodium carbonate samples are titrated with hydro-chloric acid and the total basicity is determined as percent sodium oxide or sodium carbonate (Note 1).

PROCEDURE

1. Dry the sample at 110°C for two hours.
2. Weigh out three 0.25 to 0.35 g samples into 250 ml flasks and dissolve in about 100 ml distilled water.
3. Add three drops of modified methyl orange to each sample and titrate with standard hydrochloric acid to the gray end point observed in Experiment 7A.
4. Calculate the percent sodium oxide, Na_2O, using 30.99 as the equivalent weight or percent sodium carbonate using 53.00 as the equivalent weight.

Note 1. This type sample is suitable for a glass electrode pH titration. See Experiment 10 and Chapter 8 for details.

Kjeldahl Determination of Nitrogen

Principle: The Kjeldahl method is used for the determination of ammonium and amide nitrogen. Organic amide bonds are
catalytically converted into ammonium ions. The ammonium ions are neutralized to ammonia and the ammonia
is distilled into a boric acid solution which is then titrated
with standard hydrochloric acid. The method is particularly suited for nitrogen analysis of protein materials.

PROCEDURE

1. Assemble a Kjeldahl distillation apparatus as shown in Figure
   Exp. 9-1. It consists of a 500 ml Kjeldahl flask, a trap to prevent droplets of sodium hydroxide solution from being carried
   over into the condenser, a condenser, an adaptor, and a receiver
   flask.
2. Weigh three 1 g samples onto a small filter paper. Wrap the

   sample in the paper and transfer each to a dry Kjeldahl flask
   (Note 1).
3. Add 10 g of potassium sulfate and 1.0 g powdered copper sulfate
    to each flask (Note 2).
4. Add 25 ml of concentrated sulfuric acid, clamp the flask in an in-
    clined position in a well ventilated hood. Heat the contents gently
    with a small flame. The mixture will froth and turn black at first.
    When the frothing ceases, increase the flame until the solution
    boils gently. Continue the digestion until the solution becomes
    colorless or very light yellow (Note 3).
5. Remove the source of heat, allow the flask to cool, carefully add
    200 ml of distilled water, and swirl.
6. Cool the solution under tap water to room temperature. Add 1 g
    of granulated zinc (Note 4).
7. Position a 250 ml flask containing 50 ml of 5% boric acid as a
    receiver. The tip of the adaptor should be about 1/4 inch below
    the surface of the acid (Notes 5 and 6).
8. With the spray trap connected to the condenser and the receiver
    in place, carefully pour 50 ml of cold 50% sodium hydroxide solu-
    tion down the side of the flask such that the two layers do not
    mix.
9. Connect the Kjeldahl flask to the spray bulb and swirl the flask to
    mix the solution. Immediately bring the solution to a boil and
    heat until about 100 ml of distillate is collected.
10. Remove the flame and disconnect the adaptor without allowing
    the solution to be sucked back into the distilling flask from the
    receiver. Rinse the inside and outside of the adaptor into the
    receiver.
11. Add three drops of bromcresol green indicator and titrate with
    $0.1N$ hydrochloric acid until the color changes from blue to
    lemon-yellow.
12. Calculate the percent nitrogen in the sample using 14.01 as the
    equivalent weight (Note 7).

Note 1. This insures that no sample clings to the wall of the flask.
Note 2. The potassium sulfate is used to raise the boiling point of
           the solution which increases the oxidation rate of the sample.
           The copper sulfate serves as a catalyst. Other catalysts which
           may be used are mercuric oxide, selenium dioxide, or copper
           selenite.

Note 3. The digestion may take two to three hours. If much acid is lost by evaporation, additional acid should be added.

Note 4. The granulated zinc prevents bumping during distillation by the slow formation of hydrogen gas bubbles.

Note 5. Ammonia reacts with the boric acid forming ammonium borate which can be titrated with a standard solution of hydrochloric acid.

$$NH_3 + H_3BO_3 \rightleftharpoons NH_4^+ + H_2BO_3^-$$

$$H_2BO_3^- + H^+ \rightleftharpoons H_3BO_3$$

Note 6. An alternative procedure is to pipet 50.0 ml of 0.1N hydrochloric acid into the receiver. The excess hydrochloric acid is determined by titration with 0.1N sodium hydroxide.

Note 7. For highly accurate work, the student should run a blank determination which contains all the reagents except the sample.

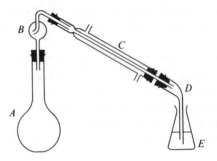

Figure Exp. 9-1

Kjeldahl distillation apparatus: (*A*) Kjeldahl flask, (*B*) trap, (*C*) condenser, (*D*) adaptor, (*E*) receiver.

Experiment 10

Glass Electrode Titration—Determination of K_a and Equivalent Weight

Principle: A solution of a weak acid in water is titrated with standard sodium hydroxide, the pH of the solution being plotted as a function of added sodium hydroxide. The equivalent weight of the acid is determined from the volume of base required to neutralize the acid. The K_a(s) is determined from the pH at various stages in the titration.

PROCEDURE

1. Obtain your sample and consult the instructor for instructions as to drying.
2. Familiarize yourself with the operations of the pH meter you will use. Electrodes are somewhat fragile, and expensive, so handle carefully. Always place the meter on off or standby before removing electrodes from the solution. They should be kept in distilled water when not in use.
3. Standardize the pH meter by immersing the electrodes in a refer-

ence buffer solution of a known pH and adjusting the meter to read the correct pH.

4. Weigh one 0.5 g sample of weak acid into a 250 ml beaker. Dissolve it in approximately 50 ml distilled water (Note 1).

5. Rinse the electrodes with distilled water and immerse them in the sample solution. Add a magnet to the beaker, place the beaker on a magnetic stirrer, and stir the solution.

6. Titrate the sample with $0.1N$ sodium hydroxide, recording pH values at 0.0 ml of titrant and each 2 ml interval thereafter until the pH becomes fairly constant above pH 10.0.

7. Plot the data just obtained with pH on the ordinate and volume on the abscissa. From this plot determine whether there is one or two breaks (one or two K_a's) and select the proper weight of sample to require approximately 40 ml of titrant.

8. Weigh out two additional samples, dissolve as before, and titrate taking readings at least 2 ml apart when not near the end point(s). Reduce the increments to 1 ml and 0.5 ml when approaching the end point, and take readings 0.1 ml apart for about 0.5 ml before and after the end point. Increase the volume increments after the end point in a symmetrical manner as they were reduced approaching the end point. Readings should be taken to at least 10 ml past the final end point. Volume readings need only be highly accurate near the end point (2 or 3 ml each side).

9. Plot the data as before, expanding the curve near the equivalence point(s), and determine the experimental end point (Note 2).

10. Calculate the equivalent weight using the following equation:

$$\text{equivalent weight} = \frac{\text{sample wt} \times 1000}{V \times N}$$

11. Calculate the K_a(s) for the acid from the pH at 25%, 50%, and 75% neutralization. Report the average of the three values. If two breaks are obtained compare the observed pH at the first end point with the value calculated by using the equation $\sqrt{K_1 K_2}$.

Note 1. It may be necessary to use an ethyl alcohol-water solvent for the sample. This will not affect the equivalent weight determination but pH measurements will be "apparent pH" readings and K_a will be an "apparent K_a" and may not agree with constants determined in water alone.

Table Exp. 10-1

pH Titration Data for an Acid-Base Neutralization

V, ml	pH	ΔV	ΔpH	ΔpH/ΔV	$\bar{V}$
35.00	6.20				
		1.00	0.20	0.20	35.50
36.00	6.40				
		1.00	0.35	0.35	36.50
37.00	6.75				
		0.50	0.37	0.74	37.25
37.50	7.12				
		0.20	0.38	1.90	37.60
37.70	7.50				
		0.10	0.35	3.50	37.75
37.80	7.85				
		0.10	1.37	13.70	37.85
37.90	9.22				
		0.10	0.58	5.80	37.95
38.00	9.80				
		0.10	0.35	3.50	38.05
38.10	10.15				
		0.20	0.35	1.75	38.20
38.30	10.50				
		0.20	0.15	0.75	38.40
38.50	10.65				
		0.50	0.25	0.50	38.75
39.00	10.90				
		1.00	0.30	0.30	39.50
40.00	11.20				
		1.00	0.18	0.18	40.50
41.00	11.38				

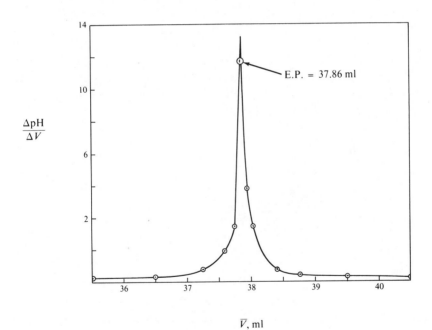

Figure Exp. 10-1

Derivative curve for pH titration.

331

Note 2. The end point can be determined by locating the midpoint of the steeply rising portion of the curve. Another graphical method, the derivative method, involves a plot of the change in pH per unit change in volume of titrant, $(\Delta pH/\Delta V)$, as a function of the volume (average) of reagent added. Typical data obtained near the end point is given in Table Exp. 10-1 and plotted in Figure Exp. 10-1. The end point is taken as the maximum and is obtained by extrapolation of the experimental points.

Oxidation-Reduction Determination
of Iron in an Acid Soluble Ore

Principle: After dissolution in hydrochloric acid, the oxidation state of the iron is adjusted to iron(II) by the addition of stannous chloride. The excess stannous chloride is oxidized with mercuric chloride. After addition of sulfuric acid and phosphoric acid, the iron is titrated with standard potassium dichromate solution using diphenylamine sulfonate as an indicator.

PROCEDURE

1. Dry the sample for one hour at 110°C, cool in a desiccator for 30 minutes, and weigh triplicate samples of 0.4 to 0.5 g (to the nearest 0.1 mg) into three 250 ml Erlenmeyer flasks.
2. Add 10 ml of distilled water and 10 ml of concentrated HCl to each flask and heat to near boiling until the sample has dissolved. No solid dark particles should remain, and a residue of white or gray silica may be ignored.

3. Dilute each sample with water to about 15 ml. From this point each sample must be treated individually to the completion of the titration.

4. Heat the solution to boiling, remove from the heat, and add $0.5M$ stannous chloride dropwise until the yellow color of Fe(III) disappears. Add two drops of stannous chloride in excess.

5. Cool to room temperature, add 50 ml of distilled water, and quickly add, all at once with swirling, 10 ml of $0.17M$ mercuric chloride. A silky white precipitate should form. If it is gray or black or if no precipitate forms within a couple of minutes, the sample must be discarded.

6. After a waiting period of two to three minutes (no longer) dilute to 100 ml, add 5 ml of concentrated sulfuric acid, 10 ml of concentrated phosphoric acid, eight drops of sodium diphenylamine sulfonate indicator, and titrate immediately with $0.10N$ potassium dichromate until the permanent appearance of a deep violet color.

7. Repeat steps 4, 5, and 6 with the remaining samples.

8. Calculate the results and report the percent iron to the nearest 0.01%.

Preparation and Standardization of Potassium Permanganate

Principle: Potassium permanganate is not a primary standard material. It contains traces of manganese dioxide. Even if pure, traces of dust and/or organic matter in the water cause reduction of some permanganate. Any manganese dioxide present in a permanganate solution catalyzes the further decomposition so it must be removed prior to standardization. After removal of the manganese dioxide the solution is standardized with sodium oxalate.

PROCEDURE

Preparation

1. Weigh 3.25 ± 0.05 g of potassium permanganate, dissolve in 1000 ml of water in a 2 liter flask, and heat just to boiling. Keep the solution about 90°C for one hour. Cover the flask with an inverted beaker and allow the solution to stand overnight so that the manganese dioxide may settle.

2. Filter the solution through a Gooch crucible with an asbestos mat or a sintered glass crucible (Note 1) into a clean filter flask. Use the first 50 ml of filtrate to rinse the filter flask and a clean glass stoppered storage bottle (Note 2).
3. Wrap the storage bottle in brown paper, tape in place, and store in the dark to prevent photochemical decomposition.

Standardization

1. Dry primary standard sodium oxalate for one hour at 110°C. Do not dry overnight.
2. Weigh three 0.25–0.32 g portions of sodium oxalate into 250 ml flasks. Dissolve in 60 ml of water and add 15 ml of 1:8 sulfuric acid.
3. Heat to 80–85°C and titrate slowly with the permanganate solution to the first permanent faint pink tinge. Stir the solution vigorously during the titration (Notes 3, 4, and 5). Do not allow the temperature to drop below 65°C during the titration.
4. Calculate the normality of the permanganate using the following equation:

$$N = \frac{\text{wt of } Na_2C_2O_4}{\dfrac{Na_2C_2O_4}{2000} \times V}$$

Note 1. The permanganate must not be filtered through paper as it will cause additional manganese dioxide to form in the filtrate.

Note 2. The filtered solution must not come in contact with rubber, cork, or any other organic matter.

Note 3. Due to the color of the permanganate the reading of the meniscus is more difficult. Either read the top of the meniscus or use a flashlight or lighted match behind the buret to illuminate the bottom of the meniscus.

Note 4. The first few drops of permanganate react very slowly. Add 0.2 to 0.4 ml and wait for it to decolorize before proceeding. Alternately one may add about 0.1 g of a manganous salt at the beginning of the titration to catalyze the reaction.

Note 5. Add the permanganate slowly near the end point to prevent error-causing side reactions from occurring.

Determination of Calcium in Limestone

Principle: This experiment illustrates an *indirect* volumetric determination of calcium. The limestone is dissolved in hydrochloric acid and the calcium ion is precipitated as calcium
oxalate which is washed, filtered, and dissolved in acid
which converts the oxalate into oxalic acid. The oxalic acid
is then titrated with standard potassium permanganate
solution. The titration determines the amount of oxalate,
which is stoichiometrically equivalent to the amount of
calcium in the original sample.

PROCEDURE

1. Dry the sample for one hour at 110°C. Weigh three 0.2 to 0.3 g
   samples into 400 ml beakers.
2. Add 10 ml of distilled water to each sample and cover the beakers
   with watch glasses. Add 10 ml of 6N hydrochloric acid from a
   pipet, allowing the acid to run down the side of the beaker into the

calcium carbonate suspension. The watch glass must be kept on during this operation to prevent loss by spattering.

3. When effervescence ceases, add a second 10 ml portion of 6N hydrochloric acid. After the samples have completely dissolved (heat can be applied if they are slow to dissolve), carefully rinse off the underside of the watch glass into its proper beaker. Dilute the volume to about 100 ml.

4. Heat the solution to near boiling and add 100 ml of hot, filtered, 5% ammonium oxalate solution.

5. Add three or four drops of methyl orange indicator and precipitate the calcium oxalate by the slow dropwise addition of 1:1 ammonia solution until the solution turns yellowish-pink (pH 3.5–4.5) (Note 1).

6. Allow the solution to stand for 30 minutes at about 80°C, then filter through a sintered glass filter. Wash the precipitate with 8 to 10 small portions of ice cold water (Note 2).

7. Remove the crucibles from the filtering assembly and place each in the beaker in which the precipitate was formed.

8. Add 50 ml of hot (80°C) 1:8 sulfuric acid so that it covers the precipitate in the crucible. Heat the solution to about 80°C and titrate the liberated oxalic acid with standard potassium permanganate solution, leaving the crucible in the beaker (Note 3), and following the precautions taken in Experiment 12.

9. Calculate the percent calcium oxide or calcium carbonate in the sample.

Note 1. Do not stop at this point. Plan your work so that you have sufficient time to complete step 6 before the end of the period. Prolonged digestion will cause an appreciable precipitation of magnesium oxalate, if magnesium is present. Because the calcium oxalate is precipitated from a solution in which the solubility is high, a well formed crystalline precipitate, having a minimum of coprecipitation, is obtained.

Note 2. Restrict the washings because of the appreciable solubility of calcium oxalate.

Note 3. Do not add the permanganate too rapidly into the sintered glass crucible without mixing with the rest of solution in the beaker. If this precaution is not observed MnO_2 may be formed, because the reaction of permanganate with oxalate to form manganous ions requires acid and the acid may be consumed in a localized area such as in the crucible.

Experiment **14**

Preparation of Standard Iodine Solutions and the Determination of Antimony in Stibnite

Principle: A titration of a reducing agent with iodine is called *iodimetry*. A titration with a reducing agent of iodine liberated in a reaction is called *iodometry*. This experiment is an example of the former, the latter will be illustrated in Experiment 15. In this experiment an iodine solution is prepared and standardized against arsenious oxide. It is then used to determine the antimony content of stibnite. Stibnite is the common antimony ore containing antimony sulfide, silica, and small amounts of other contaminates. The sample is decomposed in hot, concentrated hydrochloric acid which causes the sulfide to be evolved as hydrogen sulfide. Potassium chloride is added to prevent the loss of volatile antimony trichloride. The antimony(III) in the sample is titrated to antimony(V) with a standard iodine solution using starch indicator:

$$H_3SbO_3 + I_2 + H_2O \rightleftharpoons H_3SbO_4 + 2I^- + 2H^+$$

339

The iodine solution is standardized against pure arsenious oxide:

$$H_3AsO_3 + I_2 + H_2O \rightleftharpoons H_3AsO_4 + 2I^- + 2H^+$$

PROCEDURE

Preparation of $0.1N$ I_2 solution (Note 1)

1. Weigh about 40 g of reagent grade potassium iodide in a beaker and dissolve in about 25 ml of distilled water. Add 12.7 g of reagent grade iodine and stir until solution is complete.
2. Transfer the solution to a clean *glass-stoppered* bottle and dilute to 1 liter.
3. Store the solution in a cool, dark place (Note 2).

Standardization of $0.1N$ I_2 solution

1. Dry primary standard grade arsenious oxide at 110°C for one hour.
2. Weigh 0.2 g samples into 250 ml flasks. Dissolve in 10 ml of $1N$ sodium hydroxide solution.
3. Add 15 ml of $1N$ sulfuric acid, then carefully neutralize the excess acid with small portions of solid sodium carbonate. Add about 2 g excess to form a buffer.
4. Add 5 ml of starch indicator (Note 3) and titrate with the iodine (triiodide) solution to the first faint blue or purple color that persists for at least 30 seconds.
5. Calculate the normality of the iodine solution by the following equation:

$$N = \frac{\text{wt of As}_2O_3}{\dfrac{As_2O_3}{4000} \times V}$$

Analysis of stibnite ore

1. Dry the sample for one hour at 110°C.
2. Weigh sufficient sized samples to consume 25–40 ml of standard iodine solution into 250 ml flasks.

3. Place the samples in a hood and add 0.3 g of solid potassium chloride (Note 4) and 10 ml of concentrated hydrochloric acid.
4. Cover each flask with a small watch glass and heat to just below boiling until only a white or slightly gray residue of silica remains (Note 5).
5. Add 3 g of solid tartaric acid (Note 6) plus 5 ml of 6N hydrochloric acid, and heat for another 10 to 15 minutes.
6. While swirling the solution, slowly add distilled water until the volume is about 100 ml (Notes 7 and 8).
7. Add three drops phenolphthalein indicator, and 6N sodium hydroxide until the first pink color is obtained. Add 6N hydrochloric acid drop by drop until the pink color disappears, and then one ml excess.
8. Add 3 to 4 grams of solid sodium bicarbonate, 5 ml of starch solution, and titrate with standard iodine solution to the first blue or purple color which persists for at least 30 seconds.
9. Calculate the percentage of antimony in the sample by the following equation:

$$\%Sb = \frac{V_{I_2} \times N_{I_2} \times \frac{Sb}{2000} \times 100}{\text{sample wt}}$$

Note 1. Iodine is relatively insoluble in water, however it reacts readily with iodide ion forming the soluble triiodide ion which behaves chemically as does iodine.

$$I_2 + I^- \rightarrow I_3^-$$
$$I_3^- + 2e^- \rightleftharpoons 3I^-$$

Note 2. It is advisable to allow the solution to stand for two or three days before standardizing.
Note 3. *Preparation of starch indicator.* Make a paste of 2 g soluble starch and 10 mg mercuric iodide (preservative) in a small amount of water. Add the suspension slowly to 1 liter of boiling water. Continue boiling until the solution is clear. Cool and transfer to a glass-stoppered bottle.
Note 4. This forms the nonvolatile chloride complxes, $SbCl_4^-$ and $SbCl_6^{-3}$.

Note 5.   Do not allow the solution to go to dryness or antimony trichloride will be lost. Add more hydrochloric acid if necessary. Hydrogen sulfide is evolved as long as ore is being dissolved. When no more hydrogen sulfide is present in the vapor the dissolution is complete.

Note 6.   Tartaric acid prevents the precipitation of antimony oxide chloride (SbOCl) as the solution is neutralized. The soluble tartrate complex, $SbOC_4H_4O_6^-$, is readily and completely oxidized by iodine so no error results.

Note 7.   Too rapid addition of water causes the formation of white SbOCl which is slow to redissolve.

Note 8.   If reddish Sb_2S_3 forms, stop the addition of water, add more hydrochloric acid, and reheat.

Experiment 15

The Iodometric Determination
of Copper in an Ore

Principle: An ore containing copper is brought into solution with nitric acid. An excess of iodide is added which reacts with the copper(II) to give insoluble copper(I) iodide and iodine:

$$2Cu^{++} + 4I^- \rightleftharpoons 2CuI + I_2 \qquad \textbf{(15-1)}$$

The reaction is run under conditions such that equilibrium lies quantitatively to the right. Normal interferences are iron, arsenic, and antimony. Interference from iron is prevented by converting it into the highly undissociated complex ion FeF_6^{-3}. Arsenic and antimony must be converted to the $+5$ oxidation state where they do not interfere if the pH is greater than about 3.5. In order for Equation (15-1) to be quantitative the pH must be less than about 4. Therefore, ammonium bifluoride is added as both a complexing agent and a buffer. The iodine which is liberated in Equation (15-1) is titrated with a standard solution of sodium thiosulfate:

343

$$I_2 + 2S_2O_3^= \rightleftharpoons S_4O_6^= + 2I^- \qquad (15\text{-}2)$$

Pure potassium iodate, in a slightly acid solution, reacts with iodide ion to form iodine:

$$IO_3^- + 5I^- + 6H^+ \rightleftharpoons 3I_2 + 3H_2O \qquad (15\text{-}3)$$

The iodine released in Equation (15-3) is used to standardize the sodium thiosulfate solution as per Equation (15-2).

PROCEDURE

Preparation of $0.1N$ sodium thiosulfate

1. Heat 1 liter of distilled water to boiling in a beaker covered with a watch glass. Boil for five minutes (Note 1).
2. Cool, add about 25 g of $Na_2S_2O_3 \cdot 5H_2O$ and 0.1 g sodium carbonate (Note 2). Stir until solution is complete and transfer to a glass-stoppered bottle.
3. Allow the solution to stand overnight before standardizing (Note 3).

Standardization of $0.1N$ sodium thiosulfate

1. Dry a sample of pure potassium iodate for one hour at 110°C. Cool, weigh 0.12–0.15 g samples into 250 ml flasks and dissolve in 25 ml distilled water.
2. Add 2 g of iodate-free potassium iodide (Note 4), 10 ml of 1N hydrochloric acid, and titrate *immediately* with the thiosulfate solution. When the color of the solution reaches a pale straw-yellow, add 5 ml of starch indicator and titrate to the disappearance of the blue color.
3. Calculate the normality of the thiosulfate by the following equation:

$$N = \frac{\text{wt of } KIO_3}{\dfrac{KIO_3}{6000} \times V_{S_2O_3^=}}$$

Analysis of copper ore

1. Dry the sample for two hours at 110°C, cool, and weigh appropriate sized samples (about 1 g for 10–30% copper ores) into 150 ml

beakers. Add 20 ml of concentrated nitric acid and heat until all the copper is in solution.

2. Evaporate the solution to 5 ml, add 25 ml of distilled water, and boil to bring all soluble salts into solution and to expel the oxides of nitrogen (Note 5).

3. Filter the solution through a small filter paper, collecting the filtrate in a 250 ml flask. (If the residue is small and light in color, no filtration is necessary). Wash the residue thoroughly with hot 1:100 nitric acid.

4. Evaporate the solution to about 25 ml, cool, and add 1:1 ammonia slowly to the first appearance of the deep blue ammonia complex.

5. Add 2.0 g ammonium bifluoride and swirl until completely dissolved.

6. Add 3 g of potassium iodide and titrate with 0.1N thiosulfate to a light yellow color. Add 5 ml of starch indicator and continue until the blue color has almost disappeared, then add 2 g of potassium thiocyanate (Note 6) and continue the titration until the blue color disappears. The blue color should not return within 10 or 15 minutes (Note 7).

7. Calculate the percent copper by the following equation:

$$\%Cu = \frac{V_{S_2O_3^=} \times N_{S_2O_3^=} \times \dfrac{Cu}{1000} \times 100}{\text{sample wt}}$$

Note 1. Boiling the water destroys bacteria which metabolize the thiosulfate ion, converting it to sulfite, sulfate, and elemental sulfur.

Note 2. Sodium carbonate is added as a preservative; thiosulfate solutions are found to be most stable at a pH of 9 to 10.

Note 3. If a sediment forms, decant the clear liquid into a clean bottle. If elemental sulfur appears at any time, the solution should be discarded.

Note 4. To test the potassium iodide, dissolve 5 g KI in 200 ml of distilled water and add 10 ml of 6N HCl and 10 ml of starch solution. No blue color should appear within 10 minutes.

Note 5. If the ore is not readily dissolved by nitric acid, add 5 ml of concentrated hydrochloric acid and heat until only a small white or grey residue remains. Do not evaporate to dryness. Cool, add 10 ml of concentrated sulfuric acid, and evaporate in the hood until copious white fumes of sulfur trioxide are observed. Cool, carefully add 15 ml of distilled water and

10 ml of saturated bromine water. Boil the solution until all the bromine has been removed. Cool and proceed as in step 3 above.

Note 6. Copper(I) iodide tends to adsorb a small amount of iodine on its surface. The addition of potassium thiocyanate causes the adsorbed iodine to be released to the solution.

Note 7. A quick return of the color may be caused by the addition of too much ammonia, too little bifluoride, or the presence of oxides of nitrogen.

Fajans Determination of Chloride

Principle: The chloride in the sample is titrated with standard silver nitrate in the presence of an adsorption indicator. The indicator used is the sodium salt of dichlorofluorescein. Prior to the equivalence point, chloride ions are in excess and are adsorbed as primary ions which in turn adsorb sodium ions as counter ions. Just past the equivalence point silver ions are in excess and are adsorbed as primary ions which now adsorb the dichlorofluoresceinate ion as counter ions causing the surface of the precipitate to turn pink. Dextrin is added to prevent coagulation of the precipitate.

PROCEDURE

1. Dry the sample for at least two hours at 110°C. Cool for 30 minutes in a desiccator and weigh three 0.25–0.30 g samples into 250 ml flasks.

2. Dissolve the samples in 50 ml of chloride-free distilled water, add 10 ml of 1% dextrin solution, 10 drops of dichlorofluorescein indicator, and titrate with $0.1N$ $AgNO_3$ until the suspension turns the first permanent pink flush.

3. Calculate the percent chloride to the nearest 0.01%.

Volhard Determination of Chloride (Swift Modification)

Principle: A small but known amount of potassium thiocyanate is added to the sample and the chloride is titrated with standard silver nitrate. A high concentration of ferric ions are added to complex the thiocyanate so that it will not precipitate until the chloride has been precipitated. This procedure eliminates the need for filtration (or coating of the silver chloride) and back-titration required by the conventional Volhard procedure.

PROCEDURE

1. Dry the chloride sample for at least two hours at 110°C. Cool the sample for 30 minutes in a desiccator and weigh triplicate samples of 0.25 to 0.30 g (to the nearest 0.1 mg) into three 250 ml Erlenmeyer flasks.
2. Dissolve the sample in 20 ml of chloride-free distilled water. Add 20 ml of $6M$ HNO_3 and 10 ml of $2M$ $Fe(NO_3)_3$. Pipet exactly 1.00 ml of standard $0.01N$ KSCN (Note 1) into the mixture.

3. Titrate the solution with standard $0.1N$ AgNO$_3$ until the red color due to FeSCN^{++} just disappears (Note 2).
4. Subtract the milliequivalents of potassium thiocyanate from the milliequivalents of silver nitrate and use this figure to calculate the percent chloride in the sample to the nearest 0.01%.

Note 1.   The $0.01N$ KSCN can conveniently be made by diluting 10.00 ml of standard $0.1000N$ KSCN to exactly 100.0 ml.

Note 2.   The disappearance of the red-colored FeSCN^{++} is not as easy to detect as is the appearance of the red color in conventional Volhard titrations due to the slight yellow color of iron(III) species in solution. If uncertain of the end point, take a buret reading and then add an additional drop of silver nitrate. If no change occurs the previous reading is taken as the end point. If a definite change occurs, record this reading and repeat the operation.

Preparation and Standardization of EDTA Solutions—Determination of Zinc by Complexometric Titration with EDTA

Principle: An EDTA (ethylenediaminetetraacetic acid) solution is standardized against pure calcium carbonate. Zinc in an aqueous solution is titrated with EDTA in an ammonia-ammonium ion buffer using Eriochrome Black T or Calmagite indicator.

PROCEDURE

Preparation of solutions

1. EDTA, 0.01 M. Dissolve about 3.8 g of the dihydrate ($Na_2H_2Y \cdot 2H_2O$) in 1 liter of distilled water. Store in a plastic bottle.
2. Buffer, pH 10. Dissolve 32 g of ammonium chloride in 285 ml of concentrated ammonia and dilute to 500 ml with distilled water.
3. Standard calcium solution. Weigh into a 100 ml beaker approximately 0.5 g (to the nearest 0.0001 g) of pure calcium carbonate which has been dried at 110°C for two hours and cooled in the

desiccator. Add 10 ml of distilled water and then dissolve by the dropwise addition of 11 ml of 6*N* hydrochloric acid using a watch glass to prevent loss of sample. When dissolution is complete, transfer quantitatively to a 100 ml volumetric flask and dilute to the mark.

Standardization of EDTA solution

1. Pipet 10 ml aliquots of the standard calcium solution into three 250 ml flasks (Note 1).
2. Pipet 10 ml aliquots of a 0.02% magnesium chloride solution into the calcium solution (Note 2).
3. Add 6 ml of pH 10 buffer and Eriochrome Black T or Calmagite indicator (Note 3) and titrate immediately with EDTA solution until the indicator changes from red to a pure sky blue with no trace of red.
4. Run a blank on a solution containing 1 ml of 6*N* hydrochloric acid and 10.0 ml of 0.02% magnesium chloride. Add buffer and indicator and titrate as in step 3.
5. Subtract the volume of the blank from the titration volume to obtain the net volume. Use this net volume to calculate the molarity of the EDTA solution.

$$ M = \frac{\text{wt of CaCO}_3}{\dfrac{\text{CaCO}_3}{1000} \times V_{(net)} \times 10} $$

Determination of zinc

1. Receive your sample in a 100 ml volumetric flask and dilute to the mark with distilled water.
2. Pipet 25.0 ml aliquots into 250 ml flasks, add 15 ml of distilled water, 10 ml of pH 10 buffer, and Eriochrome Black T or Calmagite indicator.
3. Titrate with standard EDTA solution until the red color changes to a pure blue.
4. Calculate the number of millimoles of zinc present in the total sample.

Note 1. All glassware with which EDTA solutions will be used should be thoroughly cleaned and rinsed.

Note 2. Magnesium ions are necessary for the end point to coincide with the equivalence point.

Note 3. Calmagite is added as a solid from a self-measuring dispenser. Eriochrome Black T can be added as a solution (three or four drops) prepared by dissolving 0.2 g of the solid in a solution containing 15 ml of ethanolamine and 5 ml of ethyl alcohol.

Determination of Total Water Hardness

Principle: Calcium and magnesium salts in water form precipitates with soap. Calcium and magnesium carbonates cause a residue known as boiler scale when hard water is heated. The sum of calcium plus magnesium in water is determined by titration with the disodium salt of ethylenediaminetetraacetic acid (EDTA). The titration is performed at pH 10 in an ammonia-ammonium chloride buffer. Some magnesium is necessary to obtain a sharp end point. If no magnesium is present in the natural water supply, magnesium-EDTA chelate must be added.

PROCEDURE

1. Measure 100 ml samples (Note 1) of natural water into three clean 250 ml flasks.
2. Add 10 drops of 0.005 M magnesium-EDTA chelate, 2 ml of ammonia buffer, a small amount of Calmagite indicator to the sample,

and titrate with $0.01 M$ EDTA to a clear blue end point (Notes 2 and 3).

3. Calculate the parts per million calcium carbonate (Note 4) in the water from the equation

$$\text{ppm } CaCO_3 = V_{EDTA} \times M_{EDTA} \times CaCO_3 \times 10 \qquad \text{(Note 5)}$$

Note 1.   A larger or smaller sample may be required depending on the hardness of the water.

Note 2.   The color change occurs slowly so EDTA must be added slowly near the end point. This can be alleviated by warming the solutions to about 60°.

Note 3.   If the color change is to a violet and not to a clear blue, a high level of iron may be responsible. This may be avoided by adding a small amount of potassium cyanide *after* the buffer has been added. Do not allow this solution to subsequently come into contact with an acidic solution.

Note 4.   The calculation converts any $MgCO_3$ into its equivalent in $CaCO_3$.

Note 5. $V_{EDTA} \times M_{EDTA}$ gives millimoles of EDTA
millimoles EDTA $=$ millimoles $CaCO_3$
weight of $CaCO_3$ in milligrams $=$ millimoles $CaCO_3 \times CaCO_3$
The final 10 is needed because a 100 ml sample was used.
Parts per million equals milligrams per liter.

Spectrophotometric Determination
of Manganese in Steel

Principle: Manganese in steel is determined by dissolving the steel, oxidizing the manganese(II) to manganese(VII) and measuring the color spectrophotometrically. The steel is dissolved in $6M$ nitric acid and ammonium persulfate (ammonium peroxydisulfate) is added to oxidize the carbon. Potassium metaperiodate is added to oxidize the manganous ion to permanganate according to the following equation:

$$2Mn^{++} + 5IO_4^- + 3H_2O \rightarrow 2MnO_4^- + 5IO_3^- + 6H^+$$

Phosphoric acid is added to form a colorless complex with iron.

PROCEDURE

1. Weigh duplicate 1 g samples of standard steel and unknown steel (Note 1) into four 400 ml beakers. Add 50 ml of $6M$ nitric acid

and heat in a hood to dissolve. Boil the sample for one or two minutes after complete dissolution (Note 2).

2. Add 1 g of ammonium persulfate carefully to each sample and boil for 15 minutes to oxidize carbon and destroy excess persulfate.

3. If a brown turbidity appears due to hydrous manganese dioxide, add a few crystals of sodium sulfite, heat until the solution is clear and then boil for two minutes to expel the sulfur dioxide.

4. Dilute each solution to about 100 ml with distilled water and add 15 ml of 85% phosphoric acid to complex the iron(III) so that it does not absorb visible radiation.

5. Add 0.5 g of potassium periodate and boil for about three minutes to oxidize the manganese to permanganate. Cool the solutions below boiling, add an additional 0.1 g of potassium periodate, and boil another minute or two.

6. Cool the solution, transfer quantitatively to a 500 ml volumetric flask, mix well, and dilute to the mark with distilled water.

7. Determine the wave length of peak absorbance with one of the standard samples using distilled water as a reference (Note 3). (Check the range 480–550 mm.)

8. Measure the absorbance of each solution at the wavelength selected. Rinse the sample cell several times with each solution before reading its absorbance.

9. Calculate the molar absorptivity of permanganate for each standard steel sample and average them. Use Beer's law to calculate the concentration of permanganate in the solution from the unknown steel samples and then calculate the percent manganese in the sample. Report the result to three significant figures.

Note 1. If a standard steel sample is not available, the following procedure will suffice: Weigh 0.025 to 0.04 g manganous carbonate (to nearest 0.0001 g) into a 250 ml Erlenmeyer flask. Add about 25 ml distilled water and slowly add 5 ml concentrated sulfuric acid. When the sample is dissolved and effervescence has ceased, add 1 g ammonium persulfate and boil gently for 15 minutes. Dilute to about 100 ml, add 15 ml of 85% phosphoric acid and 0.5 g of potassium periodate. Boil gently for about three minutes. Allow to cool slightly, add an additional 0.1 g of potassium periodate. Boil for one or two minutes more. Cool the solution to room temperature. Add 2.0 ml of 2.0M iron(III) nitrate for each 1 g of

steel used in your unknown.   Dilute to 1 liter with distilled
water and measure the absorbance.   Calculate the molar absorptivity of permanganate and use it to calculate the percent
manganese in the sample.

Note 2.   If the sample is not readily attacked, dilute the nitric acid
with about 50 ml of distilled water, and continue heating.   If
the sample still does not dissolve, add 5 ml of concentrated
phosphoric acid and heat.   Some silica may remain.

Note 3.   Familiarize yourself with the correct operation of the spectrophotometer.   Check the 0 and 100% T readings at each
wavelength.

Spectrophotometric Determination
of Iron With 1,10-Phenanthroline

Principle: The iron content of a solution is determined spectro-photometrically by measurement of the absorption of the complex $[(C_{12}H_8N_2)_3Fe]^{++}$ formed between iron(II) and 1,10-phenanthroline. The complex has a very high molar absorptivity (11,100) at 508 nm. The intensity of the color is independent of pH in the range 2 to 9. The complex is stable and the color intensity does not change appreciably over very long periods of time. The pH is adjusted by the addition of sodium acetate. The iron is reduced to iron(II) with hydroxylamine hydrochloride before the color is developed.

$$2Fe^{+++} + 2NH_2OH + 2OH^- \rightleftharpoons 2Fe^{++} + N_2 + 4H_2O$$

PROCEDURE

1. Prepare a standard iron solution as follows: Weigh accurately about 0.07 g of pure ferrous ammonium sulfate, dissolve in water,

and transfer the solution to a 1 liter volumetric flask. Add 2.5 ml of concentrated sulfuric acid and dilute the solution to the mark. Calculate the concentration of the solution in mg of iron per liter (ppm).

2. Dissolve 0.1 g of 1,10-phenanthroline monohydrate in 100 ml of distilled water, warming if necessary.
3. Prepare solutions for a calibration curve by pipeting 0, 1, 5, 10, 25, and 50 ml portions of the standard iron solutions into six 100 ml volumetric flasks. Place a measured volume of unknown into two other flasks (Note 1). To each flask add 1 ml of 10% aqueous hydroxylamine hydrochloride, 10 ml of the 1,10-phenanthroline solution, and 8 ml of 10% sodium acetate solution. Dilute all solutions to the mark and allow to stand for 10 minutes.
4. Using the reagent blank prepared in step 3 as a reference, measure the absorption spectra of the sample containing 25 ml of standard iron (Note 2). Select the proper wavelength to use for the determination of iron with 1,10-phenanthroline.
5. Using the selected wavelength, measure the absorbance of each of the standard solutions and unknowns. Plot the absorbance vs. the concentration of the standards. Note whether Beer's law is obeyed.
6. From the absorbance of the unknown solutions, calculate the concentration (mg/liter) of iron in the original sample.

Note 1. Pipet 10.0 ml aliquots of sample for analysis. If the absorbance is too low, use 25.0 ml samples, if the absorbance is too high, dilute the sample accordingly.

Note 2. If a nonrecording spectrophotometer is used, take absorbance readings about 20 nm apart except in the region of maximum absorbance where intervals of 5 nm are used. Plot the absorbance vs. wavelength and connect the points to form a *smooth curve.*

Photometric Titration of Iron with EDTA

Principle: Iron(III) forms an intense violet complex with salicylic acid. At a pH of 1.7 to 2.4, EDTA forms a nearly colorless complex with iron(III) which is much stronger than the iron(III)-salicylate complex. The gradual disappearance of the violet color is not distinct enough for a good visual end point but the spectrophotometric end point is very sharp at the wavelength of maximum absorbance (about 525 nm) for the iron(III)-salicylate complex.

PROCEDURE

1. Prepare and standardize a $0.01M$ EDTA solution as in Experiment 18.
2. Prepare a solution containing 6 g of salicylic acid in 100 ml of methanol.
3. Prepare a sodium acetate-hydrochloric acid buffer by adding sufficient $1N$ hydrochloric acid to 250 ml of $1N$ sodium acetate solution to give a pH of 2.2 (use a pH meter).

4. Receive your iron sample in a 100 ml volumetric flask and dilute to the mark with distilled water.

5. Determine the proper wavelength for titration. Pipet a 2 ml aliquot of your sample into a 250 ml volumetric flask. Add 2 ml of 6% salicylic acid in methanol and 1 ml sodium acetate-hydrochloric acid buffer. Dilute to volume. Determine the absorption spectrum over the wavelength interval 350–650 nm using distilled water as a reference. Read the absorbance at 20 nm intervals except in the vicinity of the maxima in the spectrum where the readings should be at 5 nm intervals.

6. Add EDTA solution to a portion of the iron(III)-salicylate solution until the violet color disappears, and then 1 to 2 ml more. Determine the absorption spectrum for this solution. Plot the absorption spectra for both solutions on a single graph. The optimum wavelength for carrying out the photometric titration is that at which the difference between the absorptivities is a maximum.

7. Titration of sample. Pipet 10 ml aliquots (Note 1) of the original iron solution into 250 ml beakers, dilute to 50 ml with distilled water, and add 1 ml of 6% salicylic acid in methanol and 5 ml of sodium acetate-hydrochloric acid buffer. Add $0.01 M$ EDTA from a buret until the color begins to fade. Remove a portion of the solution from the beaker and measure the absorbance, which should be in the range 0.5 to 0.8. Return the solution to the beaker. Add 0.5 ml portions of EDTA, reading the absorbance after each addition until three or four readings have been taken before and after the end point.

8. Plot the absorbance vs. volume of EDTA added. Determine the end point from the intersection of the linear portions.

9. Calculate the number of millimoles of iron present in your sample.

Note 1. Smaller or larger aliquots of sample may be required. The titration volume should be 30 to 40 ml.

Appendix A   The Literature of Analytical Chemistry

The *primary* source of most new information is published research papers in chemical journals. After publication, brief abstracts of original papers are published in certain abstract journals. This enables the analytical chemist to scan the new literature for material of interest. Secondary sources of new material can be found in review articles, monographs, treatises, and textbooks both general and specialized.

A selected list of literature sources in analytical chemistry is given below. This is to acquaint the reader with available material and to provide suggestions for further reading.

ABSTRACTS

Chemical Abstracts. Published biweekly by the American Chemical Society with decennial indexes.

Chemisches Zentralblatt. German publication, most useful for period prior to publication of *Chemical Abstracts,* 1830–1907.

Analytical Abstracts. British publication, more specialized than the above.

Chemical Titles. Published by the American Chemical Society.

ANALYTICAL JOURNALS

Analyst (British)

Analytical Chemistry (American Chemical Society)

Analytica Chimica Acta (International)

Chimie Analytique (French)

Chemist-Analyst (United States)

Journal of Chromatography (International)

Journal of Electroanalytical Chemistry (International)

Journal of Gas Chromatography (International)

Microkimica Acta (Austrian)

Talanta (International)

Zeitschrift für Analytisch Chemie (German)

Zhurnal Analiticheskoi (Russian; English translation available)

REVIEWS

The most comprehensive source of analytical reviews is published annually in April by *Analytical Chemistry*. In even-numbered years *Fundamental Reviews* is published covering 30–40 different areas of analytical chemistry including references to all significant research published in the two-year period. Each area is reviewed by an expert(s) in that field. During odd-numbered years *Applications* is published which includes 16 areas of application such as air pollution, clinical chemistry, pesticide residues, and water analysis. These reviews are also written by experts in their fields.

TREATISES

Kolthoff, I.M., Elving, P.J., and Sandell, E.B., Editors, *Treatise on Analytical Chemistry,* Interscience Publishers, John Wiley & Sons, Inc., New York, 1959. A multivolume work.

Meites, L., Editor, *Handbook of Analytical Chemistry,* McGraw-Hill, Inc., New York, 1961.

Welcher, F.J., *Organic Analytical Reagents,* (4 volumes), D. Van Nostrand Company, Inc., New York, 1947.

Wilson, C.L., and Wilson, D.W., Editors, *Comprehensive Analytical Chemistry,* Elsevier Publishing Company, Inc., Amsterdam, 1959. A multivolume work.

TEXTBOOKS

General

Furman, N.H., Editor, *Scott's Standard Methods of Chemical Analysis,* vol 1 and 2, 5th ed., D. Van Nostrand Company, Inc., New York, 1959.

Hillebrand, W.F., Lundell, G.E.F., Bright, H.A., and Hoffman, J.I., *Applied Inorganic Analysis,* 2nd ed., John Wiley & Sons, Inc., New York, 1953.

Kolthoff, I.M., Sandell, E.B., Meehan, E.J., and Bruckenstein, S., *Quantitative Chemical Analysis,* 4th ed., The Macmillan Company, New York, 1969.

Laitinen, H.A., *Chemical Analysis,* McGraw-Hill, Inc., New York, 1960.

Meites, L., and Thomas, H.C., *Advanced Analytical Chemistry,* McGraw-Hill, Inc., New York, 1958.

Vogel, A.I., *Quantitative Inorganic Analysis,* 3rd ed., John Wiley & Sons, Inc., New York, 1961.

Willard, H.H., and Diehl, H., *Advanced Quantitative Analysis,* D. Van Nostrand Company, Inc., New York, 1943.

Instrumental Analysis

Bair, E.J., *Introduction to Chemical Instrumentation,* McGraw-Hill, Inc., New York, 1962.

Delahay, P., *Instrumental Analysis,* The Macmillan Company, New York, 1957.

Ewing, G., *Instrumental Methods of Chemical Analysis,* 3rd ed., McGraw-Hill, Inc., New York, 1969.

Reilley, C.N., and Sawyer, D.T., *Experiments for Instrumental Methods* McGraw-Hill, Inc., New York, 1961.

Strobel, H., *Chemical Instrumentation,* 2nd ed., Addison-Wesley Publishing Company, Inc., Boston, 1973.

Willard, H.H., Merritt, L.L., Jr., and Dean, J.A., *Instrumental Methods of Analysis,* 4th ed., D. Van Nostrand Company, Inc., Princeton, N.J., 1965.

Specific Topics

Ashworth, M.R.F., *Titrimetric Organic Analysis* (2 parts), Interscience Publishers, John Wiley & Sons, Inc., New York, 1965.

Bates, R.G., *Determination of pH: Theory and Practice,* John Wiley & Sons, Inc., New York, 1964.

Bauman, R.P., *Absorption Spectroscopy,* John Wiley & Sons, Inc., New York, 1962.

Berg, E.W., *Physical and Chemical Methods of Separation,* McGraw-Hill, Inc., New York, 1963.

Boltz, D.F., Editor, *Colorimetric Determination of Nonmetals,* John Wiley & Sons, Inc., New York, 1958.

Critchfield, F.E., *Organic Functional Group Analysis,* The Macmillan Company, New York, 1962.

Dean, J.A., *Flame Photometry,* McGraw-Hill, Inc., New York, 1960.

Duval, C., *Inorganic Thermogravimetric Analysis,* 2nd ed., Elsevier Publishing Company, Inc., Amsterdam, 1963.

Feigl, F., *Spot Tests in Inorganic Analysis,* 5th ed., American Elsevier Publishing Company, Inc., New York, 1962.

Fritz, J.S., and Hammond, G.S., *Quantitative Organic Analysis,* John Wiley & Sons, Inc., New York, 1957.

Gordon, L., Salutsky, M.L., and Willard, H.H., *Precipitation from Homogeneous Solution,* John Wiley & Sons, Inc., New York, 1959.

Heftmann, E., *Chromatography,* 2nd ed., Reinhold Publishing Corporation, New York, 1967.

Helfferich, F., *Ion Exchange,* McGraw-Hill, Inc., New York, 1962.

Koltholf, I.M., and Lingane, J.J., *Polarography,* 2nd ed., Interscience Publishers, John Wiley & Sons, Inc., New York, 1952.

Koltholf, I.M., and Stenger, V.A., *Volumetric Analysis,* 2nd ed. (2 volumes), Interscience Publishers, John Wiley & Sons, Inc., New York, 1942 and 1957.

Latimer, W.M., *The Oxidation States of the Elements and Their Potentials in Aqueous Solutions,* 2nd ed., Prentice-Hall, Inc., Englewood Cliffs, N.J., 1952.

Lingane, J.J., *Electroanalytical Chemistry,* 2nd ed., Interscience Publishers, John Wiley & Sons, Inc., New York, 1958.

Meites, L., *Polarographic Techniques,* 2nd ed., Interscience Publishers, John Wiley & Sons, Inc., New York, 1965.

Mellon, M.G., Editor, *Analytical Absorption Spectroscopy,* John Wiley & Sons, Inc., New York, 1950.

Morrison, G.H., and Freiser, H., *Solvent Extraction in Analytical Chemistry,* John Wiley & Sons, Inc., New York, 1957.

Samuelson, O., *Ion Exchangers in Analytical Chemistry,* 2nd ed., John Wiley & Sons, Inc., New York, 1963.

Sandell, E. B., *Colorimetric Determination of Traces of Metals,* 3rd ed., Interscience Publishers, John Wiley & Sons, Inc., New York, 1959.

Schwarzenbach, G., and Flaska, H., *Complexometric Titrations,* 2nd English ed. translated by H.M.N.H. Irving, Methuen & Co., Ltd., London, 1969.

Siggia, S., *Quantitative Organic Analysis via Functional Groups,* 3rd ed., John Wiley & Sons, Inc., New York, 1963.

Snell, F.D., and Snell, C.T., *Colorimetric Methods of Analysis* (7 volumes), D. Van Nostrand Company, Inc., New York, 1948–71.

Welcher, F.J., *The Analytical Uses of Ethylenediaminetetraacetic Acid,* D. Van Nostrand Company, Inc., Princeton, N. J., 1957.

Yoe, J.H., and Koch, H.J., *Trace Analysis,* John Wiley & Sons, Inc., New York, 1957.

Appendix B   Tables

Table B-1

Solubility Product Constants

Compound	Formula	K_{sp}
Aluminum hydroxide	$Al(OH)_3$	2×10^{-32}
Barium carbonate	$BaCO_3$	8×10^{-9}
Barium chromate	$BaCrO_4$	2.4×10^{-10}
Barium fluoride	BaF_2	1.7×10^{-6}
Barium iodate	$Ba(IO_3)_2$	1.5×10^{-9}
Barium oxalate	BaC_2O_4	2.3×10^{-8}
Barium sulfate	$BaSO_4$	1.1×10^{-10}
Cadmium carbonate	$CdCO_3$	3×10^{-14}
Cadmium oxalate	CdC_2O_4	1.5×10^{-8}
Cadmium sulfide	CdS	8×10^{-27}
Calcium carbonate	$CaCO_3$	5×10^{-9}
Calcium fluoride	CaF_2	4×10^{-11}
Calcium hydroxide	$Ca(OH)_2$	5.5×10^{-6}
Calcium oxalate	CaC_2O_4	2×10^{-9}
Calcium phosphate	$Ca_3(PO_4)_2$	1×10^{-26}
Calcium sulfate	$CaSO_4$	6×10^{-5}
Chromium hydroxide	$Cr(OH)_3$	6.0×10^{-31}
Copper(I) iodide	CuI	5.1×10^{-12}
Copper(I) thiocyanate	$CuSCN$	4×10^{-14}
Copper(II) sulfide	CuS	9×10^{-36}
Iron(II) hydroxide	$Fe(OH)_2$	8×10^{-16}
Iron(III) hydroxide	$Fe(OH)_3$	4×10^{-38}
Lead bromide	$PbBr_2$	3.9×10^{-5}
Lead chloride	$PbCl_2$	1.6×10^{-5}
Lead chromate	$PbCrO_4$	1.8×10^{-14}
Lead fluoride	PbF_2	4×10^{-8}
Lead hydroxide	$Pb(OH)_2$	1.2×10^{-15}
Lead iodide	PbI_2	7.1×10^{-9}
Lead sulfate	$PbSO_4$	1.6×10^{-8}
Lead sulfide	PbS	8×10^{-28}
Magnesium ammonium phosphate	$MgNH_4PO_4$	2.5×10^{-13}
Magnesium carbonate	$MgCO_3$	1×10^{-13}
Magnesium fluoride	MgF_2	7×10^{-9}
Magnesium hydroxide	$Mg(OH)_2$	1.2×10^{-11}
Magnesium oxalate	MgC_2O_4	9×10^{-5}
Manganese hydroxide	$Mn(OH)_2$	4×10^{-14}

Table B-1 (*continued*)

Compound	Formula	K_{sp}
Mercury(I) bromide	Hg_2Br_2	5.8×10^{-23}
Mercury(I) chloride	Hg_2Cl_2	1.3×10^{-18}
Mercury(I) iodide	Hg_2I_2	4.5×10^{-29}
Silver arsenate	Ag_3AsO_4	1×10^{-22}
Silver bromate	$AgBrO_3$	6.6×10^{-5}
Silver bromide	$AgBr$	5.2×10^{-13}
Silver carbonate	Ag_2CO_3	8.2×10^{-12}
Silver chloride	$AgCl$	1.8×10^{-10}
Silver chromate	Ag_2CrO_4	2.4×10^{-12}
Silver cyanide	$AgCN$	2×10^{-12}
Silver iodate	$AgIO_3$	3×10^{-8}
Silver iodide	AgI	8.3×10^{-17}
Silver hydroxide	$AgOH$	2×10^{-8}
Silver oxalate	$Ag_2C_2O_4$	1×10^{-12}
Silver phosphate	Ag_3PO_4	1.3×10^{-20}
Silver sulfate	Ag_2SO_4	1.6×10^{-5}
Silver sulfide	Ag_2S	2×10^{-49}
Silver thiocyanate	$AgSCN$	1.1×10^{-12}
Strontium carbonate	$SrCO_3$	2×10^{-9}
Strontium oxalate	SrC_2O_4	5.6×10^{-8}
Strontium sulfate	$SrSO_4$	2.8×10^{-7}
Zinc hydroxide	$Zn(OH)_2$	2×10^{-14}
Zinc oxalate	ZnC_2O_4	7.5×10^{-9}
Zinc sulfide	ZnS	1×10^{-24}

Table B-2

Ionization Constants for Acids and Bases

ACIDS		
Acid	Formula	K_a
Acetic	CH_3COOH	1.8×10^{-5}
Arsenic	H_3AsO_4	
K_1		6.0×10^{-3}
K_2		1.1×10^{-7}
K_3		3.0×10^{-12}
Benzoic	C_6H_5COOH	6.3×10^{-5}
Boric	H_3BO_3	5.8×10^{-10}
Carbonic	H_2CO_3	
K_1		4.3×10^{-7}
K_2		4.8×10^{-11}
Chloroacetic	$CH_2ClCOOH$	1.5×10^{-3}
Dichloroacetic	$CHCl_2COOH$	5.5×10^{-2}
EDTA	H_4Y	
K_1		8.5×10^{-3}
K_2		1.8×10^{-3}
K_3		5.8×10^{-7}
K_4		4.6×10^{-11}
Formic	$HCOOH$	1.7×10^{-4}
Fumaric	trans-$HOOCCH:CHCOOH$	
K_1		9.6×10^{-4}
K_2		4.1×10^{-5}
Hydrazoic	HN_3	1.9×10^{-5}
Hydrocyanic	HCN	4.9×10^{-10}
Hydrofluoric	HF	6.8×10^{-4}
Hydrogen sulfide	H_2S	
K_1		5.7×10^{-8}
K_2		1.2×10^{-15}
Hypochlorous	$HClO$	3.0×10^{-8}
Iodic	HIO_3	1.6×10^{-1}
Lactic	$CH_3CHOHCOOH$	1.3×10^{-4}
Maleic	cis-$HOOCCH:CHCOOH$	
K_1		1.2×10^{-2}
K_2		6.0×10^{-7}
Malic	$HOOCCHOHCH_2COOH$	
K_1		4.0×10^{-4}
K_2		8.9×10^{-6}

Table B-2 (*continued*)

Acid	Formula	K_a
Malonic	$HOOCCH_2COOH$	
K_1		1.4×10^{-3}
K_2		2.2×10^{-6}
Mandelic	$C_6H_5CHOHCOOH$	4.3×10^{-4}
Nitrous	HNO_2	5.1×10^{-4}
Oxalic	$HOOCCOOH$	
K_1		8.8×10^{-2}
K_2		5.1×10^{-5}
Periodic	HIO_4	2.3×10^{-2}
Phenol	C_6H_5OH	1.4×10^{-10}
Phosphoric	H_3PO_4	
K_1		7.5×10^{-3}
K_2		6.2×10^{-8}
K_3		4.8×10^{-13}
Phosphorous	H_3PO_3	
K_1		7.1×10^{-3}
K_2		2.0×10^{-7}
o-Phthalic	$C_6H_4(COOH)_2$	
K_1		1.2×10^{-3}
K_2		3.9×10^{-6}
Propionic	CH_3CH_2COOH	1.3×10^{-5}
Salicylic	$C_6H_4(OH)COOH$	1.05×10^{-3}
Sulfuric	H_2SO_4	
K_1		strong
K_2		1.2×10^{-2}
Sulfurous	H_2SO_3	
K_1		1.3×10^{-2}
K_2		6.3×10^{-8}
Tartaric	$HOOCCHOHCHOHCOOH$	
K_1		9.1×10^{-4}
K_2		4.3×10^{-5}
Trichloroacetic	CCl_3COOH	2.2×10^{-1}

BASES

Base	Formula	K_b
Ammonia	NH_3	1.8×10^{-5}
Aniline	$C_6H_5NH_2$	4.0×10^{-10}
Butylamine	$C_4H_9NH_2$	4.1×10^{-4}

Table B-2 (*continued*)

Acid	Formula	K_b
Ethanolamine	$HOCH_2CH_2NH_2$	2.8×10^{-5}
Ethylamine	$CH_3CH_2NH_2$	5.4×10^{-4}
Ethylenediamine	$H_2NCH_2CH_2NH_2$	
K_1		1.28×10^{-4}
K_2		2.0×10^{-7}
Hydrazine	NH_2NH_2	
K_1		1.0×10^{-6}
K_2		1.3×10^{-6}
Hydroxlamine	$HONH_2$	1.1×10^{-8}
Methylamine	CH_3NH_2	4.4×10^{-4}
Piperidine	$C_5H_{11}N$	1.6×10^{-3}
Pyridine	C_6H_5N	1.5×10^{-9}
Triethanolamine	$(HOCH_2CH_2)_3N$	6.6×10^{-7}
Tris(hydroxymethyl)- aminomethane	$(HOCH_2)_3CNH_2$	1.2×10^{-6}
Urea	NH_2CONH_2	1.5×10^{-14}

Table B-3

Standard Electrode Potentials*

Half-reaction	E°, volts
$F_2 + 2H^+ + 2e = 2HF$	3.06
$S_2O_8^= + 2e = 2SO_4^=$	2.01
$Co^{+++} + e = Co^{++}$	1.82
$H_2O_2 + 2H^+ + 2e = 2H_2O$	1.77
$MnO_4^- + 4H^+ + 3e = MnO_2 + 2H_2O$	1.695
$PbO_2 + SO_4^= + 4H^+ + 2e = PbSO_4 + 2H_2O$	1.685
$Ce^{+4} + e = Ce^{+3}$	1.61
$BrO_3^- + 6H^+ + 5e = \frac{1}{2}Br_2 + 3H_2O$	1.52
$MnO_4^- + 8H^+ + 5e = Mn^{++} + 4H_2O$	1.51
$Mn^{+3} + e = Mn^{++}$	1.51
$Au^{+3} + 3e = Au$	1.50
$PbO_2 + 4H^+ + 2e = Pb^{++} + 2H_2O$	1.455
$Cl_2 + 2e = 2Cl^-$	1.42
$Cr_2O_7^= + 14H^+ + 6e = 2Cr^{+3} + 7H_2O$	1.33
$Tl^{+3} + 2e = Tl^+$	1.25
$MnO_2 + 4H^+ + 2e = Mn^{++} + 2H_2O$	1.23
$O_2 + 4H^+ + 4e = 2H_2O$	1.229
$IO_3^- + 6H^+ + 5e = \frac{1}{2}I_2 + 3H_2O$	1.195
$Cu^{++} + 2CN^- + e = Cu(CN)_2^-$	1.12
$Br_2(l) + 2e = 2Br^-$	1.0652
$V(OH)_4^+ + 2H^+ + e = VO^{++} + 3H_2O$	1.00
$AuCl_4^- + 3e = Au + 4Cl^-$	1.00
$Pd^{++} + 2e = Pd$	0.987
$NO_3^- + 3H^+ + 2e = HNO_2 + H_2O$	0.94
$2Hg^{++} + 2e = Hg_2^{++}$	0.920
$Cu^{++} + I^- + e = CuI$	0.86
$Ag^+ + e = Ag$	0.799
$Hg_2^{++} + 2e = 2Hg$	0.789
$Fe^{+++} + e = Fe^{++}$	0.771
$PtCl_4^= + 2e = Pt + 4Cl^-$	0.68
$Ag_2SO_4 + 2e = 2Ag + SO_4^=$	0.653
$MnO_4^- + e = MnO_4^=$	0.564
$H_3AsO_4 + 2H^+ + 2e = H_3AsO_3 + 2H_2O$	0.559
$I_2 + 2e = 2I^-$	0.535
$Cu^+ + e = Cu$	0.521
$O_2 + 2H_2O + 4e = 4OH^-$	0.401
$Ag(NH_3)_2^+ + e = Ag + 2NH_3$	0.373

*Most of the values given here are taken from W. M. Latimer, *The Oxidation States of the Elements and Their Potentials in Aqueous Solutions,* 2nd ed., Prentice-Hall, Inc., New York, 1952.

Table B-3 (*continued*)

Half-reaction	E°, volts
$VO^{++} + 2H^+ + e = V^{+3} + H_2O$	0.361
$Fe(CN)_6^{-3} + e = Fe(CN)_6^{-4}$	0.36
$Ag_2O + H_2O + 2e = 2Ag + 2OH^-$	0.344
$Cu^{++} + 2e = Cu$	0.337
$AgCl + e = Ag + Cl^-$	0.222
$Cu^{++} + e = Cu^+$	0.153
$Sb_2O_3 + 6H^+ + 6e = 2Sb + 3H_2O$	0.152
$Sn^{+4} + 2e = Sn^{++}$	0.15
$S + 2H^+ + 2e = H_2S$	0.141
$UO_2^{++} + e = UO_2^+$	0.05
$2H^+ + 2e = H_2$	0.000
$HgI_4^{-4} + 2e = Hg + 4I^-$	−0.04
$Pb^{++} + 2e = Pb$	−0.126
$Sn^{++} + 2e = Sn$	−0.136
$AgI + e = Ag + I^-$	−0.151
$Ni^{++} + 2e = Ni$	−0.250
$V^{+3} + e = V^{++}$	−0.255
$PbCl_2 + 2e = Pb + 2Cl^-$	−0.268
$Co^{++} + 2e = Co$	−0.277
$Tl^+ + e = Tl$	−0.3363
$In^{+3} + 3e = In$	−0.342
$Cd^{++} + 2e = Cd$	−0.403
$Cr^{+3} + e = Cr^{++}$	−0.41
$Fe^{++} + 2e = Fe$	−0.440
$S + 2e = S^=$	−0.48
$Ga^{+3} + 3e = Ga$	−0.53
$Ni(OH)_2 + 2e = Ni + 2OH^-$	−0.72
$Cr^{+3} + 3e = Cr$	−0.74
$Zn^{++} + 2e = Zn$	−0.763
$Zn(NH_3)_4^{++} + 2e = Zn + 4NH_3$	−1.03
$Mn^{++} + 2e = Mn$	−1.18
$ZnO_2^= + 2H_2O + 2e = Zn + 4OH^-$	−1.216
$Zn(CN)_4^= + 2e = Zn + 4CN^-$	−1.26
$Mn(OH)_2 + 2e = Mn + 2OH^-$	−1.55
$Al^{+3} + 3e = Al$	−1.66
$Mg^{++} + 2e = Mg$	−2.37
$Na^+ + e = Na$	−2.714
$Ca^{++} + 2e = Ca$	−2.87
$K^+ + e = K$	−2.925
$Li^+ + e = Li$	−3.045

Table B-4

Formula Weights

AgBr	187.78	$CaSO_4$	136.14
$AgBrO_3$	235.78	$Ca_3(PO_4)_2$	310.18
AgCl	143.32	Cr_2O_3	151.99
AgI	234.77	CuI	190.44
$AgNO_3$	169.87	$Cu(NO_3)_2$	187.55
AgOH	124.88	$Cu(OH)_2$	97.55
AgSCN	165.95	CuS	95.60
Ag_2CrO_4	331.73	CuSCN	121.62
Ag_2O	231.74	$CuSO_4$	159.60
Ag_2SO_4	311.80	EDTA (H_4Y)	292.25
Ag_3PO_4	418.58	$Fe(NH_4)_2(SO_4)_2 \cdot 6H_2O$	392.14
$Al(OH)_3$	78.00	$FeCl_3$	162.21
Al_2O_3	101.96	$Fe(OH)_3$	106.87
As_2O_3	197.84	Fe_2O_3	159.69
As_2S_3	246.04	Fe_3O_4	231.54
$BaCl_2$	208.25	HBr	80.92
$BaCl_2 \cdot 2H_2O$	244.27	$HC_2H_3O_2$ (acetic acid)	60.05
$BaCO_3$	197.35	$HCO_2C_6H_5$ (benzoic	
$BaCrO_4$	253.35	acid)	122.12
$Ba(IO_3)_2$	487.15	$H_2C_2O_4$	90.04
BaO	153.34	$H_2C_2O_4 \cdot 2H_2O$	126.07
$Ba(OH)_2$	171.35	$H_2C_8H_4O_4$ (phthalic	
BaS	169.40	acid)	166.14
$BaSO_4$	233.40	HCl	36.46
CH_3OH	28.01	$HClO_4$	100.46
CO_2	44.01	HF	20.01
$C_2H_5NH_2$ (ethylamine)	45.08	HIO_4 (periodic acid)	191.91
C_2H_5OH (ethanol)	46.07	HNO_2	47.01
$(CH_3CO)_2O$ (acetic		HNO_3	63.02
anhydride)	102.09	H_2O	18.01
C_6H_5Br (bromo-		H_2O_2	34.01
benzene)	157.02	H_3PO_4	98.00
$CaCO_3$	100.09	H_2S	34.08
$CaCl_2$	110.99	H_2SO_3	82.08
CaC_2O_4	128.10	H_2SO_4	98.08
CaF_2	78.08	HSO_3NH_2	
CaO	56.08	(sulfamic acid)	97.09
$Ca(OH)_2$	73.09	$HgCl_2$	271.50

Table B-4 (*continued*)

Hg_2Cl_2	472.09	NH_2NH_2 (hydrazine)	32.05
HgO	216.59	$NaBr$	102.90
KBr	119.01	$NaCl$	58.44
KCl	74.56	NaF	41.99
KCN	65.12	$NaHCO_3$	84.01
K_2CrO_4	194.20	NaH_2PO_4	119.98
$K_2Cr_2O_7$	294.19	Na_2HPO_4	141.96
$KHC_2O_4 \cdot H_2C_2O_4$	218.16	Na_2CO_3	105.99
$KHC_8H_4O_4$ (KHP)	204.23	$Na_2C_2O_4$	134.00
KI	166.01	$Na_2H_2Y \cdot 2H_2O$ (EDTA)	372.24
KIO_3	214.00	$NaOH$	40.00
KIO_4	230.00	Na_2SO_3	126.04
$KMnO_4$	158.04	Na_2SO_4	142.04
KNO_3	101.11	$Na_2S_2O_3$	158.11
KOH	56.11	$Na_2S_2O_3 \cdot 5H_2O$	248.19
$KSCN$	97.18	P_2O_5	141.94
K_2SO_4	174.27	$PbCl_2$	278.10
Li_2CO_3	73.89	$PbCrO_4$	323.18
$MgCl_2$	95.22	PbO_2	239.19
$MgCO_3$	84.32	$PbSO_4$	303.25
$MgNH_4PO_4$	137.32	SO_2	64.06
MgO	40.31	SO_3	80.06
$Mg_2P_2O_7$	222.57	Sb_2O_3	291.50
$MnCO_3$	114.95	SiO_2	60.08
MnO_2	86.94	$SnCl_2$	189.60
NH_3	17.03	SnO_2	150.69
NH_4Cl	53.49	$SrCO_3$	147.63
$(NH_4)HF_2$	57.04	$SrSO_4$	183.68
NH_4NO_3	66.04	TiO_2	79.90
NH_4OH	35.05	V_2O_5	181.88
$(NH_4)_2S_2O_8$	228.20	$ZnCl_2$	136.28
$(NH_2)_2CO$ (urea)	60.05	$ZnNH_4PO_4$	178.38
$(NH_4)_2C_2O_4$	124.10	ZnO	81.37
NH_2OH_2		$Zn_2P_2O_7$	304.68
(hydroxylamine)	33.03		

Table B-5

Two-Place Logarithm Table

N	0	1	2	3	4	5	6	7	8	9
1	00	04	08	11	15	18	20	23	26	28
2	30	32	34	36	38	40	42	43	45	46
3	47	49	51	52	53	54	56	57	59	59
4	60	61	62	63	64	65	66	67	68	69
5	70	71	72	72	73	73	74	75	76	77
6	78	79	79	80	80	81	81	82	83	84
7	84	85	86	86	87	88	88	89	89	90
8	90	91	91	92	92	93	93	94	94	95
9	95	96	96	97	97	98	98	99	99	99

Table B-6

Properties of Concentrated Acids and Bases

	Wt.%	Density g/ml	Molarity
Acetic acid	99.7	1.05	17.4
Ammonium hydroxide (aqueous ammonia)	28	0.89	14.6
Hydrochloric acid	37	1.18	12.0
Nitric acid	70	1.40	15.6
Phosphoric acid	85	1.69	14.7
Sulfuric acid	96	1.84	18.0

Appendix C   Answers to Problems

Chapter 1

1. (a) 247.80, (b) 42.39, (c) 119.38, (d) 352.02, (e) 325.28, (f) 1359.51
2. (a) 424.68, (b) 55.28, (c) 113.45, (d) 127.08
3. 2.57
4. (a) 7.102 g (b) 3.551 g
5. 0.033 moles, $[Ba^{++}] = 0.22, [Cl^-] = 0.44$
6. (a) 2.67, (b) 0.16, (c) 0.214, (d) 0.0050, (e) 0.50, (f) 1.0×10^{-5}
7. 0.23
8. 0.025
9. 11.7
10. 0.1025
11. (a) 0.0333, (b) 0.0500
12. 0.1000
13. 1600 ml
14. $[HCOOH] = 0.19, [^-COOH] = 6.4 \times 10^{-3}$
15. 71
16. 0.0625
17. (a) 0.1, (b) 0.15, (c) 0.30, (d) 0.05
18. 0.08
19. (a) 0.91, (b) 0.35, (c) 0.84, (d) 0.23
20. 0.32

Chapter 3

1. (a) 0.0783, (b) 42.14, (c) 4.21, (d) 2.88×10^4
2. (a) 8.01, (b) 4.259, (c) 1.92, (d) 1.8
3. (a) 253, (b) 0.0183, (c) 0.315, (d) 3.858
4. (a) 0.000406, (b) 85, (c) 0.1409, (d) 20.1
5. (a) 53.23, (b) 0.15, (c) 0.05_7, (d) 1.1
6. mean, 25.5; s, 0.1_2
7. (a) 32.0640, (b) 0.0010_4, (c) 32.0640 ± 0.0013
8. 0.0990 − 0.0994
9. 90%, 0.219 − 0.225; 95%, 0.218 − 0.226
10. 0.1138 should be discarded, confidence limit, 0.1118 − 0.1126
11. 0.91
12. (a) no, (b) no
13. no
14. 5.82 can be rejected by both $\bar{d}$ and s
15. (a) absolute 0.08%, relative 11 ppt (b) absolute −0.03%, relative −4.0 ppt
16. (a) absolute 0.03%, relative 0.6 ppt (b) absolute −0.01%, relative −0.2 ppt

Chapter 4

1. 92.84
2. 9.99
3. 16.1
4. 7.29
5. 0.1 or 0.1_4
6. (a) 0.9066, (b) 0.4291, (c) 0.07673, (d) 0.1374, (e) 0.3430, (f) 0.7527
7. (a) 0.8551, (b) 0.8401
8. 0.4862
9. 8.07 g
10. 47.1 ml
11. 1.203 g
12. 11.66
13. 97.34
14. 57.45
15. 22.10
16. 0.2474 g
17. 29.2
18. 34.7% $CaCO_3$, 65.3% $BaCO_3$

Chapter 5

1. (a) $[Al^{+++}][OH^-]^3$, (b) $[Ba^{++}]^3[AsO_4^{-3}]^2$, (c) $[Ca^{++}][F^-]^2$,
   (d) $[Pb^{++}][C_2O_4^=]$, (e) $[Hg_2^{++}][Cl^-]^2$, (f) $[Tl^+]^2[S^=]$
2. (a) 1.0×10^{-16}, (b) 5.0×10^{-12}
3. 1.1×10^{-28}
4. (a) 2.0×10^{-8}, (b) 2.0×10^{-6}
5. (a) 7.2×10^{-4}, (b) 0.088, (c) $1.44 \times 10^{-3}M$
6. 2.4×10^{-8}, (b) 7.7×10^{-6}
7. 6.3×10^{-5}
8. 6.5×10^{-5}
9. (a) 1.6×10^{-18}, (b) 1.6×10^{-12}
10. (a) 1.2×10^{-3}, (b) 4.2×10^{-4}, (c) 7×10^{-5}
11. 1.3×10^{-4}
12. Will not because the product of $[A]^2[B]$ is 7.8×10^{-14} which is less than K_{sp}. (Note the dilution effect on A and B must be considered.)
13. No, 19.5% remains unprecipitated.
14. Yes, only 0.018% remains.
15. $CaCO_3$ will precipitate first. It requires $6 \times 10^{-13}M$ carbonate ion to precipitate $CaCO_3$ and $2 \times 10^{-8}M$ to precipitate $SrCO_3$, 0.003% will remain.
16. 1.5×10^{-7} for $Pb(OH)_2$ and 3.1×10^{-10} for $Cr(OH)_3$.
17. AgBr will precipitate first. It requires $2.6 \times 10^{-11}M$ silver ion to precipitate AgBr and $6.9 \times 10^{-10}M$ to precipitate AgCl. No, it is not quantitation since 3.8% remains in solution.

18. Yes, only $4.6 \times 10^{-5}\%$ of the iodide ion remains.
19. 7×10^{-6}
20. 1.4×10^{-9}

Chapter 6

1. (a) $1/1000$, (b) $1/2000$, (c) $1/2000$, (d) $1/2000$
2. 0.075%. The error would be positive.
3. 32 ml
4. 62 ml
5. 0.09999
6. 0.1051
7. 0.0957
8. 0.1369
9. $0.1435N, 0.0717M$
10. 61.14
11. 44.29
12. 0.1531
13. 84.3
14. 98.6
15. 6.02
16. 0.1772g
17. 79.4% NaCl; 20.6% KCl
18. 0.313 g
19. 97.4_6
20. 9.1% $HC_2H_3O_2$; 90.9% $(CH_3CO)_2O$
21. 40.4

Chapter 7

1. (a) 11.77, (b) 1.34, (c) 4.48
2. 12.59
3. (a) 6.3×10^{-8}, (b) 1.6×10^{-4}, (c) 2.5×10^{-12}
4. 6.60
5. 0.99
6. 13.74
7. 2.30
8. 1.24 (note if the approximate equation is used an incorrect answer of 1.17 is obtained).
9. 10.87
10. 5.95
11. 4.14
12. 5.59

13. 5.56:1
14. (a) 0.082, (b) 0.202
15. 24.3 ml
16. 8.36
17. 8.38
18. 8.00
19. 3.6
20. 5.37

Chapter 8

1. 1.48
2. 12.30
3. 6.98
6. Methyl red is best, pH at equivalence point is 5.06.
7. 1.35
8. (a) 1.83 (1.81 is obtained using Equation (8-26)) (b) 4.16, (c) 4.82, (d) 9.12, (e) 12.19
9. The first appearance of a color change occurs when 1 part of basic color appears with 10 parts of acid color. At this point the pH is 1 unit less than the pK of the indicator. Likewise the final evidence of color change occurs at a pH one unit greater than pK when one part of acid color remains with 10 parts of basic color.
10. 4.16
11. Initial pH $= 2.67$, end point pH $= 8.50$
12. (a) 1.45, (b) 2.30, (c) 4.67, (d) 7.03, (e) 9.76, (f) 12.3
13. 74.8 ml
14. Sample I contains only Na_2CO_3; Sample II contains only $NaHCO_3$; Sample III contains $NaHCO_3$ and Na_2CO_3; Sample IV contains only NaOH.
15. 90.55% Na_2CO_3, 5.65% $NaHCO_3$
16. 73.79% Na_2CO_3, 19.09% NaOH
17. 83.5% $Na_2HPO_4 \cdot 12H_2O$, 13.8% $NaH_2PO_4 \cdot H_2O$
18. 0.090M HCl, 0.169M H_3PO_4

Chapter 9

1. Coefficients are given in order of species occurring in equation (a) 1, 2, 1, 1, 2; (b) 3, 10, 6, 10, 2; (c) 4, 6, 1, 12, 6; (d) 1, 6, 6, 2, 6, 3; (e) 5, 7, 8, 7, 5, 5, 1, 8; (f) 3, 4, 18, 3, 4, 3, 3, 3; (g) 2, 27, 14, 1, 26, 30; (h) 1, 3, 8, 2, 3, 7; (i) 2, 10, 16, 2, 5, 8; (j) 1, 5, 6, 3, 3.
2. 27.32
3. $K_2Cr_2O_7 = 39.2¢$, $KMnO_4 = 30.3¢$

4. 2.4516 g
5. 43.70
6. (a) No, (b) Yes, (c) Yes, (d) No, (e) Yes
7. (b) 0.95 v, (c) 0.549 v, (e) 1.100 v
8. (a) 0.794 v, (b) 0.219 v, (c) 1.54_7 v
9. 0.571 v
10. 8.2×10^{-17}
11. (a) 1.05_5 v, (b) 0.64_2 v
12. (a) 0.407 v, (b) 0.216 v
13. 2.9×10^{-13}
14. 3.7×10^{-6}
15. -0.938 v
16. 1.5×10^5
17. 8.6×10^{55}
18. 10.0 ml = 0.735 v, 25 ml = 0.771 v, 40 ml = 0.807 v, 49 ml = 0.871 v, 49.9 ml = 0.930 v, 50.0 ml = 1.311 v, 50.1 ml = 1.384 v, 60.0 ml = 1.407 v
19. 0.826 v

Chapter 10

1. 0.0998
2. 55.51
3. 0.0986
4. 0.47:1
5. 3.5×10^{-8}
6. 2.4×10^{-12}
7. pCl = 8.14, pAg = 1.6
8. At 49.9 ml, pI = 4.00, at 50.1 ml, pI = 12.08
9. 10.0 ml pAg = 14.48; 20.0 ml pAg = 13.93; 24.0 ml pAg = 13.21; 24.9 ml pAg = 12.21; 25.0 ml pAg = 8.44; 25.1 ml pAg = 8.27; 26.0 ml pAg = 8.24; 35.0 ml pAg = 7.99; 45.0 ml pAg = 7.47; 49.0 ml pAg = 6.75; 50.0 ml pAg = 4.87; 50.1 ml pAg = 4.00; 51.0 ml pAg = 3.00; 60.0 ml pAg = 2.04

Chapter 11

1. pH 6.0 = 0.98, pH 3.0 = 0.048
2. $\dfrac{K_1}{[H^+]} + 1 + \dfrac{[H^+]}{K_2}$
3. 0.01143
4. ~ 9.7
5. Minimum pH $\sim$ 5.0. Suitable pH is 8-9
6. $0.00531 M$

7. At pH 10.00: 0 ml pCa = 1.30, 25 ml pCa = 1.78, 40 ml pCa = 2.26, 49 ml pCa = 3.30, 49.9 ml pCa = 4.30, 50.0 ml pCa = 5.90, 50.1 ml pCa = 7.50, 51 ml pCa = 8.50, 60 ml pCa = 9.46. At pH 8.00: All pCa up through 49.9 ml are the same as at pH 10.00, 50.0 ml pCa = 4.98, 50.1 ml pCa = 5.66, 51 ml pCa = 6.66, 60 ml pCa = 7.62. The inflection at the equivalence point is much sharper at pH 10.0 then at pH 8.0 therefore a titration at pH 10.0 would be better.
8. $1.3 \times 10^{-5}M$
9. (a) 5.85, (b) 7.0, (c) 3.95
10. Calcium does not form a stable enough complex with calmagite and the end point comes prematurely.
11. (a) 805, (b) 83, (c) 904
12. $Mg^{++} = 0.0221M, Zn^{++} = 0.0162M$

Chapter 12

1. $6.32 \times 10^{14} \sec^{-1}, 4.18 \times 10^{-12}$ ergs
2. (a) 0.125, (b) 0.284, (c) 1.00, (d) 1.39, (e) 2.00
3. (a) 98.9 (b) 59, (c) 33, (d) 10, (e) 2.3, (f) 1.0
4. (a) 2.4%, 1.62 (b) 68.9%, 0.162
5. 4.57×10^{-5}
6. 11,100
7. 0.84
8. 0.0981 g
9. (a) 3200 (b) 4.8 ppm or $8.2 \times 10^{-5}M$
10. 40.3%
11. 530 nm
12. acidic $= 2.11 \times 10^{-4}M$, basic $= 2.14 \times 10^{-4}M$
13. 3.2×10^{-8}
14. $A = 5.12 \times 10^{-5}M, B = 5.01 \times 10^{-5}M$
15. $Cr = 1.2 \times 10^{-3}M, Mn = 1.1 \times 10^{-4}M$
16. $A = 1.1 \times 10^{-4}M, B = 6.7 \times 10^{-5}M$
17. $0.0596M$
18. 0.0516
19. Additions of 0.50 ml to a volume of 85–90 ml do not have a significant dilution effect on the concentration (and absorbance) whereas 1.00 ml additions to a volume of 10–17 ml have a significant effect on the concentration.

Index

383

Four-Place Logarithms

N	0	1	2	3	4	5	6	7	8	9
10	0000	0043	0086	0128	0170	0212	0253	0294	0334	0374
11	0414	0453	0492	0531	0569	0607	0645	0682	0719	0755
12	0792	0828	0864	0899	0934	0969	1004	1038	1072	1106
13	1139	1173	1206	1239	1271	1303	1335	1367	1399	1430
14	1461	1492	1523	1553	1584	1614	1644	1673	1703	1732
15	1761	1790	1818	1847	1875	1903	1931	1959	1987	2014
16	2041	2068	2095	2122	2148	2175	2201	2227	2253	2279
17	2304	2330	2355	2380	2405	2430	2455	2480	2504	2529
18	2553	2577	2601	2625	2648	2672	2695	2718	2742	2765
19	2788	2810	2833	2856	2878	2900	2923	2945	2967	2989
20	3010	3032	3054	3075	3096	3118	3139	3160	3181	3201
21	3222	3243	3263	3284	3304	3324	3345	3365	3385	3404
22	3424	3444	3464	3483	3502	3522	3541	3560	3579	3598
23	3617	3636	3655	3674	3692	3711	3729	3747	3766	3784
24	3802	3820	3838	3856	3874	3892	3909	3927	3945	3962
25	3979	3997	4014	4031	4048	4065	4082	4099	4116	4133
26	4150	4166	4183	4200	4216	4232	4249	4265	4281	4298
27	4314	4330	4346	4362	4378	4393	4409	4425	4440	4456
28	4472	4487	4502	4518	4533	4548	4564	4579	4594	4609
29	4624	4639	4654	4669	4683	4698	4713	4728	4742	4757
30	4771	4786	4800	4814	4829	4843	4857	4871	4886	4900
31	4914	4928	4942	4955	4969	4983	4997	5011	5024	5038
32	5051	5065	5079	5092	5105	5119	5132	5145	5159	5172
33	5185	5198	5211	5224	5237	5250	5263	5276	5289	5302
34	5315	5328	5340	5353	5366	5378	5391	5403	5416	5428
35	5441	5453	5465	5478	5490	5502	5514	5527	5539	5551
36	5563	5575	5587	5599	5611	5623	5635	5647	5658	5670
37	5682	5694	5705	5717	5729	5740	5752	5763	5775	5786
38	5798	5809	5821	5832	5843	5855	5866	5877	5888	5899
39	5911	5922	5933	5944	5955	5966	5977	5988	5999	6010
40	6021	6031	6042	6053	6064	6075	6085	6096	6107	6117
41	6128	6138	6149	6160	6170	6180	6191	6201	6212	6222
42	6232	6243	6253	6263	6274	6284	6294	6304	6314	6325
43	6335	6345	6355	6365	6375	6385	6395	6405	6415	6425
44	6435	6444	6454	6464	6474	6484	6493	6503	6513	6522
45	6532	6542	6551	6561	6571	6580	6590	6599	6609	6618
46	6628	6637	6646	6656	6665	6675	6684	6693	6702	6712
47	6721	6730	6739	6749	6758	6767	6776	6785	6794	6803
48	6812	6821	6830	6839	6848	6857	6866	6875	6884	6893
49	6902	6911	6920	6928	6937	6946	6955	6964	6972	6981
50	6990	6998	7007	7016	7024	7033	7042	7050	7059	7067
51	7076	7084	7093	7101	7110	7118	7126	7135	7143	7152
52	7160	7168	7177	7185	7193	7202	7210	7218	7226	7235
53	7243	7251	7259	7267	7275	7284	7292	7300	7308	7316
54	7324	7332	7340	7348	7356	7364	7372	7380	7388	7396